# Regional Problems of Earth Remote Sensing (RPERS 2020)



E3S Web of Conferences Volume 223 (2020)

Krasnoyarsk, Russia
29 September – 2 October 2020

**Editors:**

**G.M. Tsybulski Gennady**
**M.V. Noskov**
**Yu.A. Maglinets**

ISBN: 978-1-7138-2437-4

**Printed from e-media with permission by:**

Curran Associates, Inc.
57 Morehouse Lane
Red Hook, NY 12571

**Some format issues inherent in the e-media version may also appear in this print version.**



For additional information, please contact EDP Sciences – Web of Conferences
at the address below.

EDP Sciences – Web of Conferences
17, Avenue du Hoggar
Parc d'Activité de Courtabœuf
BP 112
F-91944 Les Ulis Cedex A
France

Phone: +33 (0) 1 69 18 75 75
Fax: +33 (0) 1 69 28 84 91

contact-edps@webofconferences.org

**Additional copies of this publication are available from:**

Curran Associates, Inc.
57 Morehouse Lane
Red Hook, NY 12571 USA
Phone: 845-758-0400
Fax: 845-758-2633
Email: curran@proceedings.com
Web: www.proceedings.com

# TABLE OF CONTENTS

edp sciences

Web of Conferences

# Statement of Peer review

In submitting conference proceedings to *Web of Conferences*, the editors of the proceedings certify to the Publisher that

1. They adhere to its **Policy on Publishing Integrity** in order to safeguard good scientific practice in publishing.
2. All articles have been subjected to peer review administered by the proceedings editors.
3. Reviews have been conducted by expert referees, who have been requested to provide unbiased and constructive comments aimed, whenever possible, at improving the work.
4. Proceedings editors have taken all reasonable steps to ensure the quality of the materials they publish and their decision to accept or reject a paper for publication has been based only on the merits of the work and the relevance to the journal.

Title, date and place of the conference

Regional Problems of Earth Remote Sensing RPERS 2020, September 29-October 02, 2020

Proceedings editor(s):

Y.A. Maglinets

Date and editor's signature

20.10.2020

# Specificity of using wireless technologies in Earth remote sensing

*Michail* Kazakov[1], *Fedor* Kazakov[1], and *Natalia* Sirotinina[1]

[1] Siberian Federal University, Computer Science Department, 79 Svobodny pr., 660041 Krasnoyarsk, Russia

**Abstract.** The article discusses an approach to organizing the interaction of nodes of spacecraft onboard data transmission networks based on wireless communication technologies. The main quality indicators and requirements for traffic control in on-board networks are formulated taking into account the peculiarities of the wireless network nodes interaction. The results of modeling various algorithms of transmission medium access control for one of the common protocols for organizing wireless networks (IEEE 802.15.4) are presented..

## 1 Introduction

Systems of remote sensing (ERS) based on spacecraft are one of the priority areas for the development of science and the space industry [1]. The number and density of satellites that survey in various bands of electromagnetic spectrum with a resolution from several kilometers to several tens of centimeters is constantly growing [2-6] .

One of the main trends in the development of ERS in recent years is the use of multi- and hyperspectral optoelectronic systems operating in both visible and infrared ranges. This allows us to observe the earth's surface using hundreds of spectral channels. At the same time, this trend leads to an increase in the information performance required to process and transmit traffic that is generated by remote sensing systems on board a spacecraft. The information performance of modern spacecraft can reach 2 Gbit / s. For comparison, in devices of twenty years ago this parameter rarely exceeded tens of Mbit / s [7].

Currently, research is mainly focused on the information transmission over space-to-Earth communication channels. At the same time, development of spacecraft onboard data transmission systems taking into account the requirements of fault tolerance and speed, is also an important task [10].

One of the most promising approaches in this area is the use of wireless networks. It allows to reduce the use of cable bundles in onboard networks, the mass of which is up to 10% of the total mass of the spacecraft [8-9].

In contrast to conventional wire technologies, in wireless networks the data transmission medium is shared by all on board nodes of the spacecraft. When determining the effective bit rate, consideration must be given to the sharing of transmission medium. Depending on the wireless technologies used, the bit rate can be tens of Gbps.

[1] Corresponding author: mkazakov@sfu-kras.ru

When choosing protocols and developing algorithms for accessing the data transmission medium, it is necessary to take into account the situation when a delay-critical traffic (for example, control signals) and a large data amount from the spacecraft payload (for example, from equipment of hyperspectral optoelectronic systems) must be transmitted simultaneously.

In public information networks, this problem is solved by using various algorithms for traffic prioritization and setting quality of service parameters. At the same time, significant restrictions on the traffic parameters are imposed by the used protocols of the physical data transmission medium access.

IEEE 802.15.4 [11] is one of the standards corresponding to the specific application in the spacecraft on-board network. It specifies the protocol that defines the physical layer and the media access control algorithm. This standard defines two modes of access to the transmission medium, namely concurrent and guaranteed access. The first is based on the carrier sense multiple access with collision avoidance (CSMA / CA), which does not allow control of QoS parameters and allows packet loss. The second mode enables to reserve time intervals for transmitting information from a limited number of nodes. In this case, guaranteed access mode with specified characteristics is provided.

To evaluate the performance of this standard, a model was developed using the OMNeT simulation environment.

It is proposed to use the average delay in packet transmission and the percentage of packet loss due to denial of access to the transmission medium as the evaluation criteria.

An experimental study of the model showed a number of disadvantages that affect the possibility of using this standard in spacecrafts.

The results of experiments with the original IEEE 802.15.4 are shown in Figure 1. As you can see, percentage of packet loss begins to grow starting from the number of nodes equal to seven. By the time fourteen nodes are involved in transmission, the loss percentage becomes more than 12% and continues to grow.

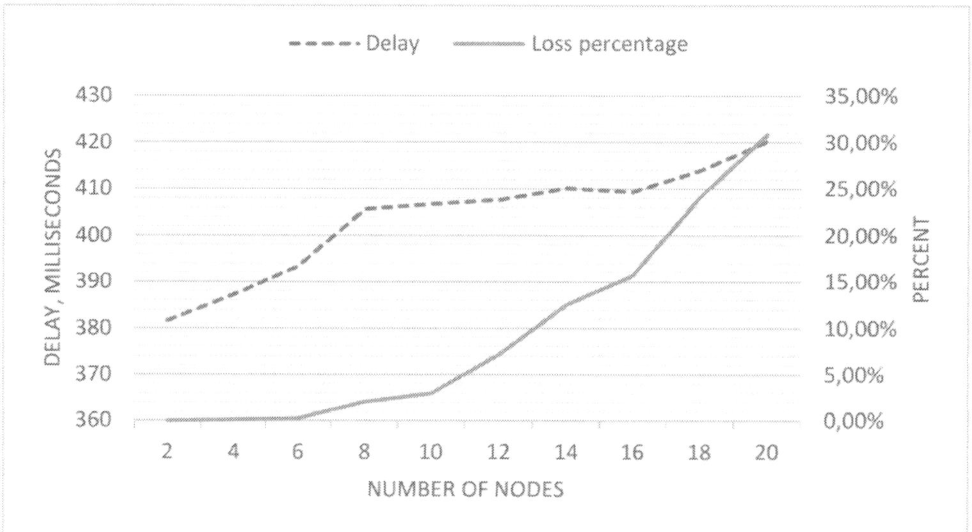


**Fig. 1.** Latency and percentage of packet loss for concurrent access

In the guaranteed access mode, a zero percentage of packet loss is provided for no more than seven nodes. The rest of the nodes cannot participate in data transmission in the guaranteed access mode.

To avoid packet loss when using the guaranteed access mode, a number of modifications of the guaranteed access slot allocation algorithm have been proposed. The modifications consist in the distribution of guaranteed slots using various algorithms. Algorithms of random selection, ring shift, priority algorithm and class prioritization were selected as the most promising.

The proposed solution made it possible to avoid the loss of packets for any number of nodes participating in the transmission. However, this increased delays in packet transmission. As you can see in fig. 2, the delay depends on the number of nodes and the algorithm used.

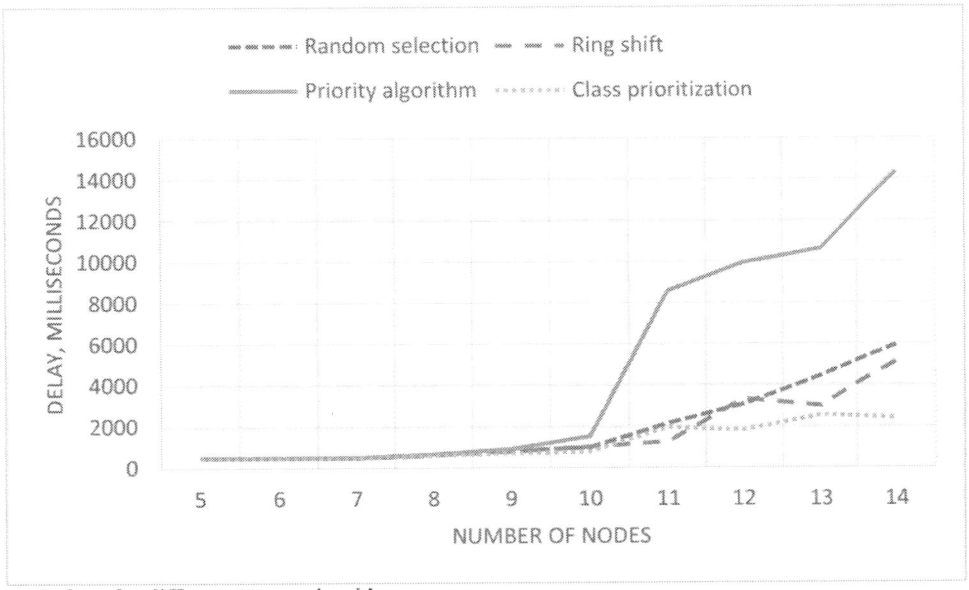


**Fig. 2.** Delays for different access algorithms

The random selection method is characterized by a graph consisting of two linearly increasing sections. The section corresponding to ten or more nodes participating in the transmission is characterized by faster growth.

For the circular shift method, the delay depends on the queue shift parameter. In general, the pattern of delay growth is similar with the random selection method. Some uneven growth is observed due to effects related to the multiplicity of the shift length to the total number of nodes.

The main way to reduce latency growth is to use a priority algorithm. In this case, different types are distinguished in the total traffic depending on the consumer category. A reasonable priority is assigned to each traffic type. Experiments have shown a significant increase in the average delay for this algorithm when the number of nodes increases over 10 (Figure 3).

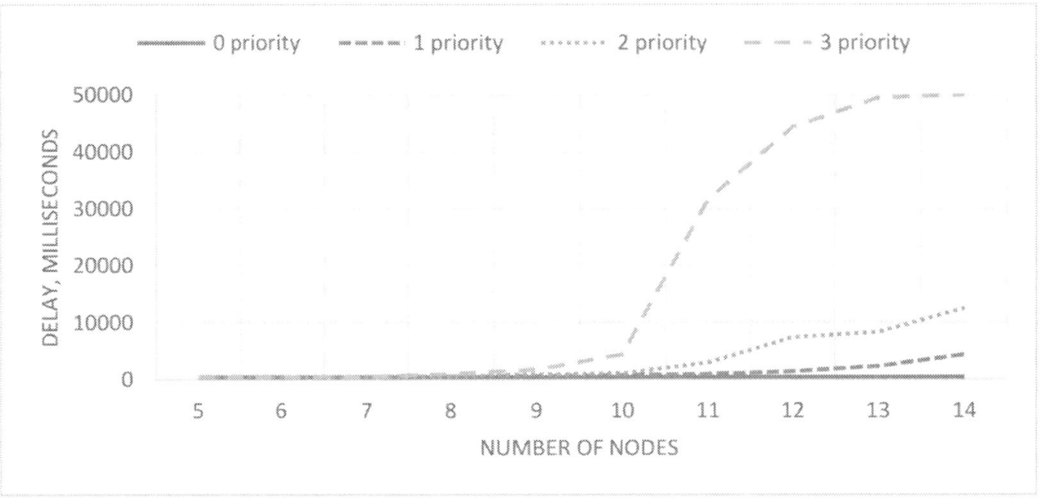


**Fig. 3.** Dependence of delay for different priorities on the number of nodes in the case of using the priority algorithm

This is due to a latency increase for low priority nodes. However, for the nodes with the highest (zero) priority, the delay time increases insignificantly (Figure 3). For nodes of the first and second priorities, the delay also increases, starting from the tenth node.

## 2 Conclusion

The article analyzes the operation of wireless sensor networks based on the IEEE 802.15.4 standard in a situation where high data transmission reliability is required and network overload is observed. The goal of the analysis is to determine the effect of number of nodes on the percentage of packet loss and latency, and to suggest approaches for improving network performance.

Simulation of the proposed modifications to distribution control of guaranteed time slots showed a complete absence of packet loss, which confirms the high reliability of data transmission even with a large number of nodes.

A side effect of this is an increase in packet transmission delay (latency) . This can be partially compensated for by using prioritized algorithms and optimizing node priorities.

The obtained results allow us to draw a conclusion about the applicability of the proposed modified protocol when building on-board wireless networks of spacecraft, including for solving remote sensing tasks.

## References

1. I.N. Gansvind, Small satellites in remote sensing of the Earth, Exploration of the Earth from Space, **5**, pp. 82-88, (2019)
2. A.V. Shpak, O.A. Zheleznyak, Geoinformation analysis of remote sensing data to forecast the possibility of forest fires, Astronomical School's Report, **7**, (2011)
3. H. LIANG, C. ZHANG, Z. MENG, Real-time observation monitoring and analysis network, ANNEX II, **2**, (2008)
4. R. Dicati, Earth Remote Sensing, Stamping the Earth from Space, pp.293-359, (2017)

5. Y.A. Maglinets, G.M. Tsibul'skii, M.V. Noskov, Current Trends of Development of the Regional Systems of Earth Remote Sensing, Sib. Fed. Univ. Eng. technol., **9(7)**, 1012-1018. (2016)
6. N.E. Zimmermann, R.A. Washington-Allen, R.D. Ramsey, M.Schaepman, Modern Remote Sensing for Environmental Monitoring of Landscape States and Trajectories, A Changing World, pp. 65-91, (2007)
7. G. N. Maltsev, I. A. Kozinov, Transfer of Hyperspectral Video Data of Earth Remote Sensing via Radio Channels of Limited Capacity, INFORMATION CONTROL SYSTEMS, **2**, pp 74-83, (2019).
8. R. A Magness, Comparison of can and bluetooth protocols a study for application of can over bluetooth for wireless on-board data handling for a spacecraft sensor network, DASIA of ESA Special Publication, (2003)
9. M. Meltzer, The Cassini-Huygens visit to Saturn: An HistoricMission to the Ringed Planet, Springer, (2015)
10. D.S. Andrahitov, S.A. Gezha, A.A. Filimonov, PROBLEM ISSUES OF INCREASING THE ACCURACY OF MEASUREMENTS ONBOARD SPACE VEHICLES, Modern High Technologies, **19**, (2020)
11. S. Li, Comparative Analysis of Infrastructure and Ad-Hoc Wireless Networks, ITM Web of Conferences, (2019)

# Technology of establishment of terrestrial objects aerospace survey-based object-oriented monitoring systems

*Yurii A.* Maglinets[1*], *Ruslan V.* Brezhnev[1]

[1] Siberian Federal University, The Institute of Space and Information Technology, Kirenskogo st. 26, Krasnoyarsk, 660074 Russia

**Abstract.** The paper studies a technology for the establishment of Earth remote sensing systems based on the object-oriented approach to the complex system analysis and design. The technology relies upon the system of knowledge combining the knowledge of a certain subject domain, the knowledge of image processing algorithms and the typology of the problems in question. Another important element of the technology is a set of human-computer interaction methods that make it possible to arrange a task setting and solving cycle without or with little involvement of image processing experts. The paper presents the input conditions and the main stages of the technology, as well as examples of possible use of the system in agricultural monitoring.

## 1 Introduction

The improving consumer quality of the ERS (earth remote sensing) data makes them more and more attractive for the end users in different domains, therefore bringing up the need for the development of hardware and software tools for the data intelligence and analysis.

By the present moment, the researcher community has developed a concept for the remote monitoring systems (RMS) [1-3] that provide the software and hardware infrastructure for special data processing. As a rule, such systems are designed for fixed situation monitoring, when the task is clearly formulated and formalized, when the scope of interest and the scope of programs and algorithms involved is clearly determined. The main downside of this approach is the lack of flexibility: any new inquiries from the users require the involvement of data intelligence programming specialists.

The key approach to changing the situation is replacing the "end user – specialist" interaction method with the "end user – solution system" in the monitoring systems [4, 5].

In the process of building of such remote monitoring systems, the researchers face the problems of diversity, dissimilarity, spatial and temporal variability of space monitoring objects. To ensure the effective interpretation of the changes that occur to a monitoring object with airspace images, the interpretation system has to be capable of operating the knowledge of its structural and behavioral properties and their relations with the spectral,

[*] Corresponding author: maglinets@mail.ru

textural, and structural properties of the analyzed scenes. One of the most promising approaches to the design of the multipurpose software and technological monitoring media is the object-oriented paradigm [6-10] that applies the object-oriented approach principles to the presentation of special data and the design of algorithms for the symbol interpretation of images. The review of the current publications brings us to the conclusion that both Russian and foreign research groups are actively studying the methods and algorithms of object-oriented monitoring manifested in the object segmentation and classification models [8-12]. The object-oriented methods are being tested and adjusted to different classes of objects including agricultural fields, water bodies, coal mining areas for the studies of dust emissions [9, 12-13] etc. The most remarkable results in the development of object-oriented monitoring methods and systems were achieved by the scientists of the Russian Space Research Institute using Vega-Science multipurpose system [13]. Besides, it should be remarked that the research groups have not yet arrived at a unified concept of the object-oriented monitoring of various special objects.

The paper presents the technology of object-oriented monitoring designed to simplify the relations between the end users and RMS by implementing the knowledge presentation and processing tools with special interactive techniques. The technology is developed at the School of Space and Information Technology of Siberian Federal University.

## 2 Technology description

The scope of the technology is the monitoring or special objects within a definite subject domain. The input condition is the presence of renewable sources of spatial data and (or) services supplying relevant information for the monitoring; the presence of the infrastructure ensuring the possibility of collection and processing of the spatial data along with the open access for the end user setting the task (hereinafter referred to as the Task Setter).

Main stages of the technology.

1. Domain ontology setting. To ensure the system's "understanding" of the user, the intrasystem presentation of the survey domain knowledge is required to limit the class of the survey domains. The authors have developed a database setting concept formulated as a system of principles invariant to the survey domain, ensuring the presentation of the thematic ontologies for the monitoring purposes [14-15]. The knowledge of the monitoring object (Object) semantics is presented as thematic ontologies in the OWL format.

Generally, an individual notion would be associated with a number of measurable properties. The lower hierarchy notions inherit the parent properties. Every measurable property is characterized with its measurement scale and the permitted value limits with the format individual for the selected scale. The property may be associated with one or more related measurement methods. The measurement method is an estimation procedure with such significant properties as the currency and correctness of the information.

2. Organization of access to the renewable spatial data base (SDB) being the region of interest for the end user. The research is based on the SDB of the SibFU School of Space and Information Technology [16] and Vega Science open web services of the Russian Space Research Institute [1]. The technology implies no constrains for the SDB presentation. It can be based on a local renewable set of vector and bitmap layers supplied with the respective attributive information or as dynamically formed sets of spatial data created with the outside web GIS services. To ensure the possibility of ontology interpretation through connecting the ontology classes and the spatial object units, the correlation between the formed domain knowledge base and the spatial object database is made to correlate every individual spatial object to an entry in the respective DBMS table that supports the presentation of such spatial relations.

3. The end user sets the task within the interface solution concept, according to which the basis for the interface organization is the abovementioned domain database and the global dialogue structure invariant to the subject domain of the survey [17]. The task-setting dialogue is carried out as a hierarchic descend on the taxonomy of the domain notions, in the process of which the user determines the target properties. The survey object properties may be used as limitations for the information search; in the given situation, the value interval shall be determined. Besides that, the target properties of the survey object to be measured in the monitoring process are determined. Based on the respective types of relations within the database, the limits for the monitoring time interval and the spatial localization of the monitoring object are set. For every measurable property, a standard pattern for its changes in time and the rules for the interpretation of possible deviations of the measurement results from the planned values may be set. A relevant parameter of the monitoring task setting is the requirement for the frequency of submission of the ERS processing results.

The paper relies upon the task-setting method developed by its authors, determining the basic steps of the dialogue (Fig. 1 *a–d*) and regulating the actions of the user and the system in the process, including: the selection of the monitoring object type and the properties of interest with the option of combining the hierarchic descend down the domain notion tree and the keyword search; the indication of the spatial location of the monitoring object by means of navigation operation, replacement of mounts, territorial division and zoning systems, as well as specification of the mentioned information in the dialogue process; setting the temporal interval for the measurement by direct or indirect methods; specification of the monitoring object model elements related to the permitted and threshold values of the measurable properties, adjustment of the policy for notification of the user of any critical situations.

4. Interpretation of the task setting by the System. In general, the task setting is considered as the System's capacity of interpreting the actions of the Task Setter in the dialogue into its intrasystem representation. The two main interpretation scenarios have been implemented and tested.

The first scenario implies that the Task Setter makes a choice within the limited multitude of previously formulated tasks (for example, the most frequently solved tasks or the tasks the System is capable of solving without additional learning). At that, the System database has the established relations of "task – solution algorithm". The variables for the previously determined algorithms can be object coordinates, points in time or date intervals as limits for the selection of data to be processed in order to solve the task.

The second scenario suggests that in according with the global dialogue structure stages the Task Setter moves down the hierarchy of the notion tree within the domain context, attributing the key properties of the monitoring object stage by stage.

This task setting method is based on resolving the monitoring task in a dialogue with the end user [18] that provides a multitude of possible actions of the user, with the System database that provides a multitude of relations of the "action – computational procedure" type.

Within the framework of the suggested technology, a model for organization of knowledge of operating spatial data in the monitoring process has been developed [18]. Unlike the models embedded in ENVI, ERDAS Imagine, IMC, ESA-SNAP and other software packages [21], the developed model makes it possible to present the knowledge of the composition, sequence, and relation of the algorithmic flowcharts of the geospatial data processing along with the relations between the spatial object, the task, the cumulation of algorithms and the temporal sequence of solving the task under prior uncertainty. The model presents the problem solving procedure as a multitude of discreet changes of the syntactic (spectral, textural) properties of the object within the given time interval; it is characterized with cyclicity, multi-tasking and distribution capacity.

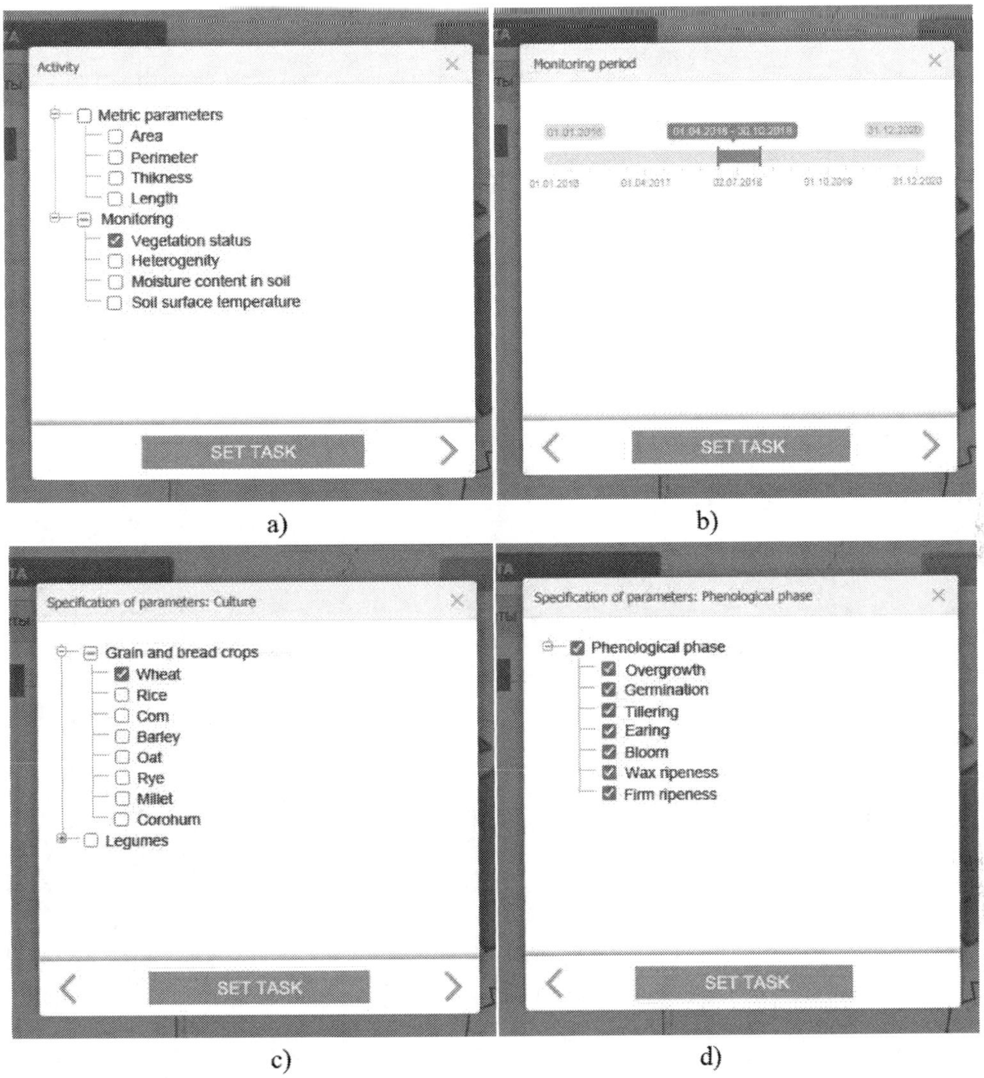


**Fig. 1.** Grain culture monitoring task-setting (for wheat)

5. The totality of actions described in paragraphs (1)-(4) is a compulsory condition for the launch of the monitoring procedure. The process is limited to the totality of measurements taken with the frequency determined by the task description. The monitoring results are presented to the end user as the triggers set in the task setting process are activated.

## 3 Implementation and testing of the technology components

The software and information components of the technology described in the present paper have been implemented in the remote monitoring system of the SibFU School of Space and Information Technology [16].

https://doi.org/10.1051/e3sconf/202022301002

In particular, the following software tools have been produced:

- The functional prototype of the monitoring task setting web interface;
- The web service for building the performed algorithmic structures for the ERS data analysis and processing;
- Set of distributed auxiliary software modules to ensure the supply of relevant aerospace data to the system, the preparation and proper processing of data, the connection of the functional blocks of algorithmic structures etc.;
- System knowledge database in the OWL format for the storage of themed references for the selected domains.

The domain selected for the technology testing was agricultural monitoring. The tasks selected for the testing of the technology components are the analysis of spatial unevenness of the agricultural circuit, the forecasting of the grain culture crops, the monitoring of the grain culture condition, and the monitoring of adherence to the agricultural technology of handling the fallow land [18-20], Fig. 2.

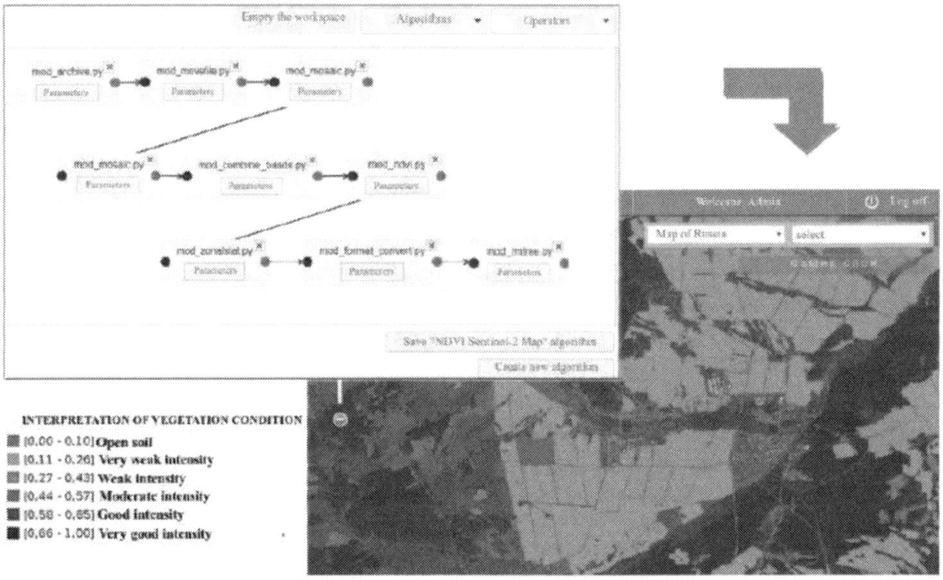


**Fig. 2.** Algorithmic structure for the monitoring of the grain culture condition and the results of one of its cycles

## 4 Conclusions

A technology of creation of the terrestrial object-oriented monitoring information systems based on aerospace survey was developed. It relies upon the object-oriented approach applied to the organization of storage of the spatial monitoring objects, the presentation of data and the knowledge of objects in relation to the temporal aspects of the alternation of their life phases and the technology workflow in order to develop some service-oriented applications based on a unified software and technological platform, including the distributed functional components intended for the collection, processing, analysis and presentation of the derived cartographic products with the capacity of adaptive adjustment of the information and computational processes of resolving the themed inquiries of the end users. The technology is intended to align the mental model of the end user on one hand with the totality of the processed data, algorithm of its adoption, presentation and processing on the other. The result of such alignment is the task solving mechanism interpreted into the intrasystem presentation. The software, information, and methodical aspect of the technology have been tested for compliance for solving the agricultural monitoring problems in the remote monitoring system of the SibFU School of Space and Information Technology.

*The reported study was funded by Russian Foundation for Basic Research (No. 18-47-242002 p_mk), Government of the Krasnoyarsk Territory, Krasnoyarsk Regional Fund of Science, to the research project: «The development of technology for creating intelligent information systems of object-oriented monitoring of territories based on remote sensing data».*

## References

1. Development of approaches to the construction of information systems for remote monitoring / E.A. Lupyan, M.A. Burtsev, A.A. Proshin, D.A. Kobets // Modern problems of remote sensing of the Earth from Space. 2018. No.3. P.53-66.
2. Features of the organization of control and management of distributed remote monitoring systems / I.V. Balashov, V.Yu. Efremov, A.A. Mazurov, Jr., A.S. Mamaev, A.M. Matveev, A.A. Proshin // Modern problems of remote sensing of the Earth from Space. 2011. Vol.8, No. 3. pp. 161-166.
3. Development of UTDIS ERS using cloud technologies / P.A. Loshkarev, O. O. Tokhiyan, A.M. Kurlykov, K.V. Koshkin, A.P. Gladkov // Geomatika. 2013. № 4. P. 22-26.
4. Creation of technologies for building information systems of remote monitoring / E.A. Lupyan, I.V. Balashov, M.A. Burtsev, V.Yu. Efremov, A.V. Kashnitskii etc. // Modern problems of remote sensing of the Earth from Space. 2015. Vol. 12. No 5. P. 53–75.
5. Traynor C., Williams M.G. End users and GIS: a demonstration is worth a thousand words // Interactive Technologies. 2001. P. 115-133.
6. Kumar, M. An Object Oriented Shared Data Model for GIS and Distributed Hydrologic Models / M. Kumar, C. Duffy // International Journal of Geographical Information Science, IJGIS-2008-0131. 2010. p. 1061–1079.
7. Pradhan, R. K. Data models and structure in Geographic Information System Application / R. K. Pradhan // International Journal of Advance Computing Technique and Applications (IJACTA), ISSN : 2321-4546. 2014. P. 40–48.
8. Gavrilyuk, E.A. Mapping of terrestrial ecosystems of the Pechora-Ilychsky nature reserve and its environs based on the reconstructed multi-temporal Landsat satellite data / E.A. Gavrilyuk, A.S. Plotnikova, D.E. Plotnikov // Regional problems of Earth remote sensing. 2018. P. 87–91.
9. Giniyatullina, O.L. Experience of using Earth remote sensing in the tasks of geoecological monitoring of a coal-mining area / O.L. Giniyatullina, E.L. Schastlivtsev, V.A. Kovalev // Regional problems of Earth remote sensing. 2018. P. 283–287.
10. Ma L., Fu T., Li M. Active learning for object-based image classification using predefined training objects // IJRS. 2018. Vol. 39, is. 9. P. 2746–2765.
11. Hossain M.D., Chen D. Segmentation for object-based image analysis (OBIA): A review of algorithms and challenges from remote sensing perspective // ISPRS J. of Photogrammetry and Remote Sensing. 2019. Vol. 150. P. 115–134.
12. Rylov, S.A. Using active learning to highlight water bodies in satellite images / S.A. Rylov // Regional problems of Earth remote sensing. 2020. P. 122-126.
13. Using the satellite monitoring service "VEGA-SCIENCE" to assess the state of agricultural crops / S.A. Linkov, A.V. Akinchin, I.S. Donchenko, A.A. Popov //

Novosti nauki v APK, 2018. No. 2(11). P. 16–20.
14. The Intelligent Managerial Decision Support System for Agricultural Land Evaluation Maglinets, Yury; Raevich, Ksenia; Tsibulsky, Gennady // E3S Web of Conf. 75, 03005 (2019).
15. Raevich K.V. Approaches to Representation of Knowledge of Operations on Spatial Data in Monitoring Applications / K.V. Raevich, Yu.A. Maglinets, R.V. Brezhnev // CEUR Workshop Proceedings. – 2020. – Vol. 2534. – P. 172-177.
16. Многоцелевая автоматизированная система космического агромониторинга ИКИТ СФУ – состояние и перспективы / R.V.Brezhnev, Yu.A. Maglinets, K.V.Raevich, G.M. Tsybulsky // Materials of the 15th All-Russian open conference "Modern problems of remote sensing of the Earth from Space" – 2017.
17. The Scheme of Setting and Solving of Spatial Objects Monitoring Tasks / Y. Maglinets, R. Brezhnev, K. Raevich, A. Pyataeva, G. Tsybulsky // IEEE, 2020 International Conference on Information Technology and Nanotechnology (ITNT), Samara, Russia, 2020, P. 1-5.
18. Brezhnev, R.V. An Interactive Environment for Modeling the Processes of ERS Data Processing and Analysis / R.V.Brezhnev, Yu.A.Maglinets, K.V.Raevich // CEUR Workshop Proceedings. – 2020. – Vol. 2534. – P. 61-67.
19. Obtaining time series of LAI to predict crop yield / E.V. Fedotova, Yu.A. Maglinets, R.V. Brezhnev, A.G. Vyrvinskiy // Modern problems of remote sensing of the Earth from Space, 2020. – Vol.17, No. 4. – P. 195–203.
20. Modeling of Agricultural Spatial Objects with Heterogeneous Dynamically Changing Spatial Structure / R.V. Brezhnev, Yu.A. Maglinets, K.V. Raevich, G.M. Tsibulski // CEUR Workshop Proceedings. 2018. Vol. 2210 P.316 -322.
21. ESA-SNAP – Stamps integrated processing for Sentinel-1 persistent scatterer interferometry / M. Foumelis, J.Manuel Delgado Blasco, Y. Desnos, M. Engdahl, D. Fernandez, L. Veci / IGARSS. 2018. P.1364-1367.

# REGIONAL MONITORING OF FORESTS USING THE VEGA-LES SYSTEM: CASE STUDY FOR TUNGUSSKO-CHUNSKOYE FOREST MANAGEMENT UNIT AND TUNGUSKA RESERVE IN THE RUSSIAN KRASNOYARSK REGION

*Alexandr* Kashnitskii*, *Ivan* Balashov, *Sergey* Bartalev, *Viacheslav* Egorov, *Dmitrii* Kobets, *Vasily* Zharko and *Evgeniy* Loupian

Space Research Institute of the RAS (IKI), 84/32 Profsoyuznaya, Moscow, Russia, 117997

**Abstract.** This paper demonstrates the capabilities of the Vega-Les ("Les" is the Russian word meaning "forest") information system (IS) for forest monitoring. A brief assessment and characteristics of the Earth observation data and main available thematic products about Russian forests available in the system are given. An assessment of the capabilities of the Vega-Les IS for studying local scale forest changes was carried out. The Tungussko-Chunskoye forest management unit (FMU) and the Tunguska nature reserve in the Russian Krasnoyarsk Krai region were chosen as the test area. The analysis of forest cover changes over this area since the beginning of the 21st century, including the changes in the number and extent of wildfires, is presented. As a result, it is concluded that the Vega-Les IS is applicable for remote assessment and monitoring of various characteristics of Russian forests.

## 1 Introduction

Vega-Les information system (IS) [1] was developed by the Space Research Institute of the Russian Academy of Sciences (IKI) in 2019 based on the "IKI-Monitoring" center for collective use [2, 3] infrastructure. The main purpose of this system is to provide researchers with the capabilities for remote complex study of forest resources in Russia using Earth observation data. The system contains a huge amount of satellite data collected over a long period of time as well as the capabilities for its analysis and processing. Current volume and composition of the main remote sensing data available in the system are presented on the "IKI-Monitoring" center website (http://ckp.geosmis.ru/default.aspx?page=6). Various regularly updated cloud free composite images produced using a fully automated satellite data processing chain are available in the system as well.

* Corresponding author: kashnizky@gmail.com

The system also includes a large number of various thematic maps of Russian forests mainly developed based on the moderate resolution data. These maps include annual land cover maps (23 classes), dominant tree species maps, growing stock volume (GSV) maps and others [4, 5, 6]. Additionally, the system contains information about forest fires and their effects. Data analysis capabilities implemented in the system include both basic geographic information system (GIS) instruments, as well as specialized information processing tools for studying the forest cover.

This paper presents the capabilities of the Vega-Les IS for the local scale analysis of the forest resources. These capabilities are demonstrated in a case study for the Tungussko-Chunskoye forest management unit (FMU) and the Tunguska reserve in the Russian Krasnoyarsk Krai region.

## 2 ANALYSIS OF FOREST COVER AND ITS CHANGES OVER THE TERRITORY OF TUNGUSSKO-CHUNSKOYE FMU AND TUNGUSSKA RESERVE

There are more than 10 million hectares of forested land on the territory of the Tungussko-Chunskoye FMU and the Tunguska reserve. The reserve is located in the center of the FMU. Its total area is 296.6 thousand hectares. The territory under consideration is characterized by a continental climate, with large variations of daily and seasonal air and soil temperatures, a small amount of atmospheric precipitation (predominantly in summer) and pronounced periods of summer drought [7]. The composition of forest stands is mainly represented by conifers - pine, larch, spruce, Siberian pine. The age of the stands is up to 200 years. There is practically no road network, the permanent population is concentrated in the administrative center and does not exceed several thousand people. Due to the vast and inaccessible territory, objective information about changes in forest cover throughout the Tungussko-Chunskoye FMU and the Tunguska reserve can only be obtained remotely. Main information about forest changes based on remote sensing data which can be obtained using Vega-Les IS is presented below.

### 2.1 Species composition of forest cover and its changes from 2006 to 2019

Estimates presented in this section are based on thematic products with information on different characteristics of forests obtained from MODIS data. These maps are regularly updated and are currently available from 2006 to 2019. A list of the maps available in the system and links to the methods for their development are given in more detail in the table 3 in work [8].

The analysis of these maps shows that larch is a dominant specie over the territory in question (total area ranges from 6.7 to 7.6 million hectares in different years). There are also pine (1.08 to 1.22 million hectares), birch (0.87 to 1.16 million hectares), Siberian pine (about 0.26 to 0.35 million hectares) and spruce (0.08 to 0.16 million hectares) forests; other dominant species are rare. To visualize annual changes in species' area let us take 2006 as the base year and calculate the changes relative to this base area. Fig. 1 shows base areas for 2006 and the annual changes in species' area.

The most significant variations in the absolute area are observed for larch. Its area decreased by almost 1 million hectares in 2017-2018. Mostly this area changed into the "recent burns" class present in the dominant tree species maps in the system, and to a lesser extent into the class with a prevalence of birch. For the remaining significant classes for this territory (pine, birch, Siberian pine) the variations in the absolute area are small, but in relative numbers they are noticeable. For other classes, including spruce, variations are not representative since the base area is too small. That being said, it can be noted that the total area of the birch-dominated forests increased by 23 percent by the end of 2019 compared to 2006. The main changes also took place in 2017-2018. The area of the classes dominated by pine and Siberian pine is slowly decreasing throughout the observation period. By 2019 their area has decreased by 12 percent for pine and by 26 percent for Siberian pine compared to the base area in 2006. Generally we associate these changes with forest fires and post-fire damage to the forest. The scale and impact of wildfires in this area is considered in the next section.

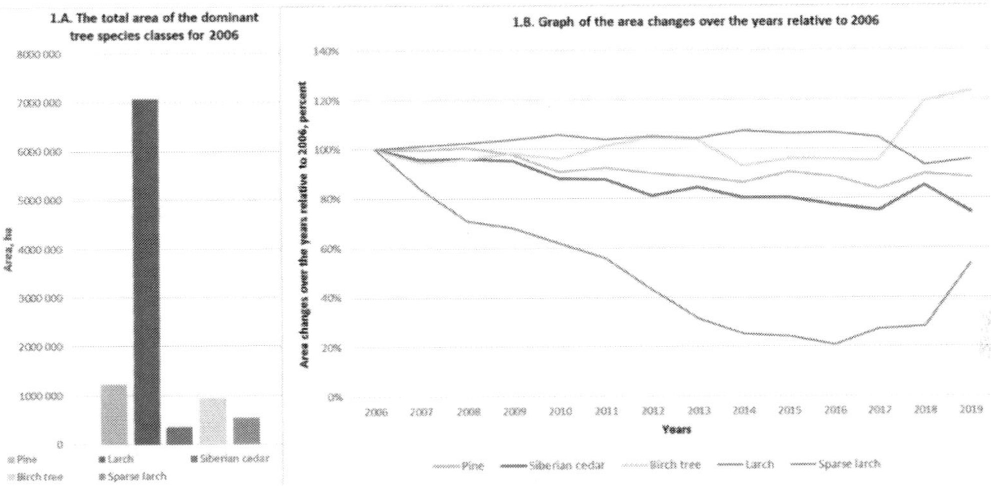


**Fig. 1.** Distribution of the area of dominant tree species. Fig. 1.A. The total area of the dominant tree species classes for 2006. Fig. 1.B. Graph of the area changes over the years relative to 2006.

## 2.2 Analysis of forest fire and post-fire damage areas

Estimates given in this section are based on the active fire detection using MODIS sensor (AQUA and TERRA satellites), as well as the improved burned area identification based on Landsat and Sentinel-2 data (ETM+, OLI and MSI sensors). The system uses a technique developed at IKI which allows combining individual "hot spots" into fires, tracking their dynamics and assessing the area. This technique is described in sufficient detail in [9, 10, 11]. The main active fire detection data was obtained for the period from 2001 to 2020 from MODIS collection 6 [12], with subsequent integration and correction according to the technique referenced above. To improve the estimates of the post-fire damage to the forest based on Landsat and Sentinel-2 data the method described in [13] was used. All 2020 fires over the FMU territory were analyzed by an expert in an interactive mode to improve the estimates of the post-fire forest damage based on Landsat and Sentinel-2 data using the abovementioned approach.

Forest fires in the area under consideration are widespread which makes extinguishing and localizing them difficult due to the inaccessibility and lack of infrastructure. The points of origin of forest fires are fairly evenly distributed over the territory in question. Fig. 2 shows maps of forest fires from 2015 to 2020 (for 2020 - until September 28) based on the indicated information from active fire detection. Fig. 3 shows the change in the number and total area of forest fires over the territory of the Tungussko-Chunskoye FMU and the Tunguska nature reserve.

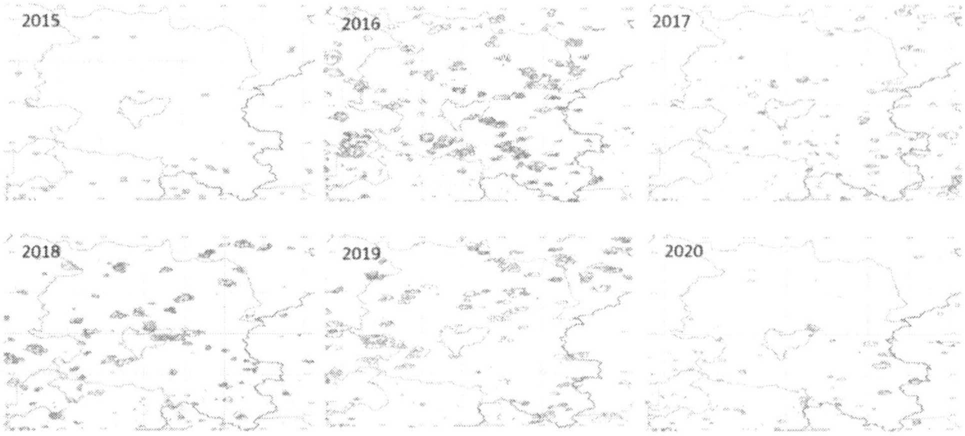


**Fig. 2.** The data from 2015 to 2020 (for 2020 - until September 28) on fires that occurred on the territory of the Tungussko-Chunskoye FMU and the Tunguska reserve.

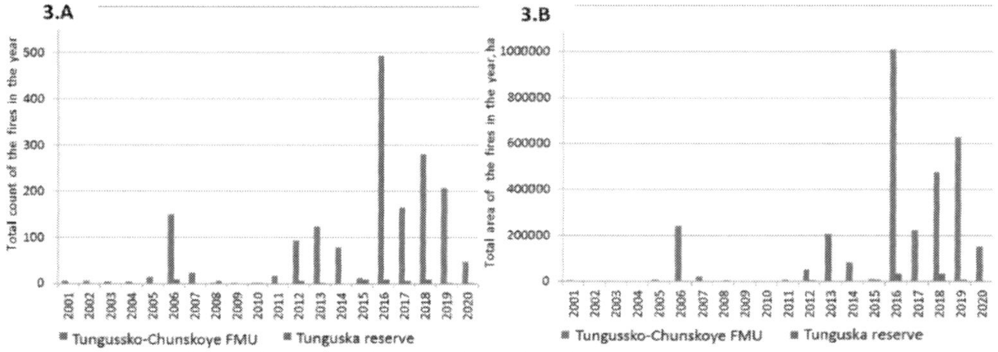


**Fig. 3.** Changes in the number and total area of forest fires in the Tungussko-Chunskoye FMU and the Tunguska fire reserve. 3.A. The total number of fires per year. 3.B. Total area covered by fires during the year.

A clear increase in the number of fires and the area covered by fires occurred in 2016-2019, with fires affecting the territory of the reserve as well. Improved estimates for 2020 based on high resolution data yielded results comparable to active fire detection data. In 2020 (at the time of September 28) 150 thousand hectares were burned according to active fire detection data, and a similar figure was obtained based on high spatial resolution data (the discrepancy was less than 1 thousand hectares). Thus the technique for assessing the area based on active fire detection gave very accurate estimates for the territory under consideration for 2020, which indicates its high reliability.

## 3 Conclusions

As a result of the analysis of data from 2001 to 2020, the following conclusions can be made about the forest cover changes in the Tungussko-Chunskoye FMU and the Tunguska reserve:

1. Fires in the area have become more intense with a visible trend towards an increase in their total number and area covered by fire. The years from 2016 to 2019 were the most intense in terms of forest fires. At the same time the current year 2020 doesn't demonstrate a high fire intensity - by the end of September the fire has covered only 150 thousand

hectares which is less than in the last four years (although more than the average at the beginning of the century).

2. The territory in question is dominated by larch. Its area in the FMU and the nature reserve is fairly constant (variations in the range from 6.7 to 7.6 million hectares). At the same time during 2017-2018 there was a sharp decrease in the area of larch. The area of other classes of dominant conifers (pine, Siberian pine) is insignificant in relation to larch. However, by the end of 2019 their total area has decreased by 20 percent compared to the maximum values for the observation period (2006-2019). Conversely the area of the class with a prevalence of birch has grown by similar amount. We attribute these changes in the species composition to fire damage to the forest.

Presented estimates are preliminary and may require some ground-based verification allowing to take into account the specifics of the area in question. At the same time, it should be noted that all the analysis capabilities and data available in the Vega-Les IS have been verified in various regions and have shown their efficiency [6]. In the future it is possible to conduct a separate study to clarify and confirm the data presented. However, the territory under consideration is characterized by extreme inaccessibility. Relevant and reliable information on it can be feasibly obtained only using remote sensing data.

Thus, based on the analysis of the territory of the Tungussko-Chunskoye FMU and the Tunguska reserve, it can be concluded that the Vega-Les IS is applicable for remote study and monitoring of various characteristics of Russian forests at the regional and local levels.

The work was carried out using the capabilities of the "IKI-Monitoring" center for collective use [1, 2] with the support of the Ministry of Science and Higher Education of the Russian Federation (theme "Monitoring", state registration No.01.20.0.2.00164).

## References

1. Balashov, I., Bartalev, S., Bartalev, S., Burtsev, M., Vorushilov, I., Egorov, V., Kashnitskii, A., Khovratovich, T., Khvostikov, S., Kobets, D., Loupian, E., Saigin, I., Senko, K., Stytsenko, F., Sychugov, I., Zharko, V. Vega-Les Information System. Actual Features and Future Evolution // IOP Conference Series: Earth and Environmental Science. IOP Publishing, Vol. 507. №. 1. P. 012002. (2020) DOI: 10.1088/1755-1315/507/1/012002
2. Loupian E.A., Proshin A.A., Bourtsev M.A., Balashov I.V., Bartalev S.A., Efremov V. Yu., Kashnitskiy A.V., Mazurov A.A., Matveev A.M., Sudneva O.A., Sychugov I.G., Tolpin V.A., Uvarov I.A. IKI center for collective use of satellite data archiving, processing and analysis systems aimed at solving the problems of environmental study and monitoring // Sovremennye problemy distantsionnogo zondirovaniya Zemli iz kosmosa. Vol. 12. №. 5. P. 263-284. (2015)
3. Loupian E.A., Proshin A.A., Bourtsev M.A., Kashnitskiy A.V., Balashov I.V., Bartalev S.A., Konstantinova A.M., Kobets D.A., Mazurov A.A., Marchenkov V.V., Matveev A.M., Radchenko M.V., Sychugov I.G., Tolpin V.A., Uvarov I.A. Experience of development and operation of the IKI-Monitoring center for collective use of systems for archiving, processing and analyzing satellite data // Sovremennye problemy distantsionnogo zondirovaniya Zemli iz kosmosa. Vol. 16. №. 3. P. 151-170. (2019) DOI: 10.21046/2070-7401-2019-16-3-151-170.
4. Bartalev S.A., Loupian E.A. R&D on Methods for Satellite Monitoring of Vegetation by Russian Academy of Sciences' Space Research Institute // Sovremennye problemy distantsionnogo zondirovaniya Zemli iz kosmosa. Vol. 10. № 1. P. 197-214. (2013)
5. Bartalev S.A., Egorov V.A., Zharko V.O., Loupian E.A., Plotnikov D.E., Khvostikov S.A. Current state and development prospects of satellite mapping methods of Russia's

vegetation cover // Sovremennye problemy distantsionnogo zondirovaniya Zemli iz kosmosa. Vol. 12. №. 5. P. 203-221. (2015)

6. Bartalev S.A., Egorov V.A., Zharko V.O., Loupian E.A., Plotnikov D.E., Khvostikov S.A., Shabanov N.V. Land cover mapping over Russia using Earth observation data // Moscow. Russian Academy of Sciences, Space Research Institute, 208 p. (2017)
7. Tunguska reserve. Biocenoses of the northern taiga and the influence of extreme natural factors on them. Proceedings of the GPZ "Tunguska". Issue 1. Tomsk: Tomsk University Publishing, 294 p. (2003)
8. Balashov I.V., Kashnitskiy A.V., Bartalev S.A., Bartalev S.S., Bourtsev M.A., Vorushilov I.I., Egorov V.A., Zharko V.O., Kobets D.A., Konstantinjva A.M., Loupian E.A., Saigin I.A., Senko K.S., Stytsenko F.V., Sychugov I.G., Khvostikov S.A., Khovratovich T.S. VEGA-Les: information system for complex monitoring of forests and hunting grounds in Russia // Sovremennye problemy distantsionnogo zondirovaniya Zemli iz kosmosa. Vol. 17. №. 4. P. 73-88. (2020) DOI: 10.21046/2070-7401-2020-17-4-73-88
9. Loupian E.A., Bartalev S.A., Balashov I.V., Egorov V.A., Ershov D.V., Kobets D.A., Senko K.S., Stytsenko F.V., Sychugov I.G. Satellite monitoring of forest fires in the 21st century in the territory of the Russian Federation (facts and figures based on active fires detection), Sovremennye problemy distantsionnogo zondirovaniya Zemli iz kosmosa, Vol. 14, №. 6, P. 158-175. (2017) DOI: 10.21046/2070-7401-2017-14-6-158-175.
10. Lupyan E.A., Mazurov A.A., Flitman E.V., Ershov D.V., Korovin G.N., Novik V.P., Abushenko N.A., Altyntsev D.A., Koshelev V.V., Tashchilin S.A., Tatarnikov A.V., Sukhinin A.I., Ponomarev E.I., Grishin A.M., Afonin S.V., Belov V.V., Gridnev Yu V.V., Matvienko G.G., Soloviev V.S., Antonov V.N., Tkachenko V.A. Satellite monitoring of forest fires in Russia. Results. Problems. Perspectives. Analytical review // IOA; GPNTB SB RAS. - Novosibirsk. (Ser. Ecology), Issue 68. 134 p. (2003)
11. Stytsenko F.V., Bartalev S.A., Ivanova A.A., Loupian E.A., Sychugov I.G. Forest burnt area assessment possibilities in regions of Russia based on active fires detection by satellites, Sovremennye problemy distantsionnogo zondirovaniya Zemli iz kosmosa, Vol. 13, №. 6, P. 289-298. (2016) DOI: 10.21046/2070-7401-2016-13-6-289-298.
12. Lyapustin, A., Wang, Y., Korkin, S., Huang, D.: MODIS Collection 6 MAIAC algorithm, Atmos. Meas. Tech., 11, 5741–5765, (2018) DOI: 10.5194/amt-11-5741-2018.
13. Kashnitskii A.V., Loupian E.A., Bartalev S.A., Bartalev S.S., Balashov I.V., Efremov V.Yu., Stytsenko F.V. Optimization of burn mapping interactive procedures in remote fire monitoring information systems // Sovremennye problemy distantsionnogo zondirovaniya Zemli iz kosmosa, Vol. 12, №. 4, P. 7-16. (2015)

# Computational Probabilistic Analysis and Transformation Method in the Earth Remote Research

*Boris* Dobronets[1,*], *Olga* Popova[1], and *Alexey* Merko[1]

[1]Institute of Space and Information Technology, Siberian Federal University, Kirenskogo 26, Krasnoyarsk, 660074 Russia

**Abstract.** The article deals with the issues of numerical modeling of problems with random input data. Finding the joint probability density function of the vector of output parameters is considered. It is proposed to use computational probabilistic analysis and the transformation method. A numerical example of the joint probability density function of the vector of a solution of a system of nonlinear equations with random input data is given.

## 1 Introduction

To solve many practical problems, including using Earth remote sensing (ERS) data, an approach based on numerical modeling is used. The issue becomes much more complicated when the input parameters of the mathematical model have various types of uncertainty, including random ones. In the case of random uncertainty of model parameters, the researcher must have knowledge of the probability density functions of random characteristics, the joint distribution function and the form of functional dependencies that establish model relationships between the input and output parameters.

The available knowledge about the properties of random parameters significantly affects the choice of methods and approaches for solving both particular problems of numerical modeling and the general problem as a whole, dictated by a practical problem.

One of the main ways to solve such problems is a numerical simulation by the Monte Carlo method. For example, this approach is used to simulate the interaction of solar radiation with the plant environment [1].

An important stage of modeling is the generation of sufficient statistics (usually tens, hundreds or more thousands) of trajectories of random parameters. Working with such a volume of information is characterized by computational complexity and low speed of convergence of the computational process.

An alternative to the Monte Carlo method is Computational Probabilistic Analysis (CPA), which offers methods for solving linear and nonlinear systems of equations with random parameters. The article discusses a mathematical model in the form of a system of functions with random arguments. A modified method is used to calculate the joint probability density function of the vector of output parameters. The approach is based on the use of CPA and the transformation method. Using the example of solving a system of nonlinear equations, an algorithm for constructing the joint density function of the solution vector is demonstrated. It uses a new concept of CPA called probabilistic extension.

*e-mail: BDobronets@yandex.ru

## 2 Transformation method

In this section, we will consider the application of the transformation method for calculating the joint probability density function of the vector of output parameters of a system of functions with random arguments [9].

Let the mathematical model be described by a system of functions

$$\begin{aligned} z_1 &= f_1(x_1, \dots, x_n), \\ &\vdots \\ z_m &= f_m(x_1, \dots, x_n), \end{aligned} \tag{1}$$

where $Z = (z_1, \dots, z_m)$, $X = (x_1, \dots, x_n)$, $m \leq n$.

The joint probability density function $p_Z$ of a random vector $Z$ can be found directly in terms of the joint probability density function $p_X$ of a random vector $X$. This will require defining the equivalent events of infinitesimal rectangles.

For example, consider the linear transformation of random vectors for $m = n$

$$Z = AX.$$

Let us denote by A the matrix of dimension $n \times n$. Suppose that the matrix $A$ has an inverse, therefore, each vector $X$ corresponds to a unique $Z$ obtained from

$$X = A^{-1}Z \tag{2}$$

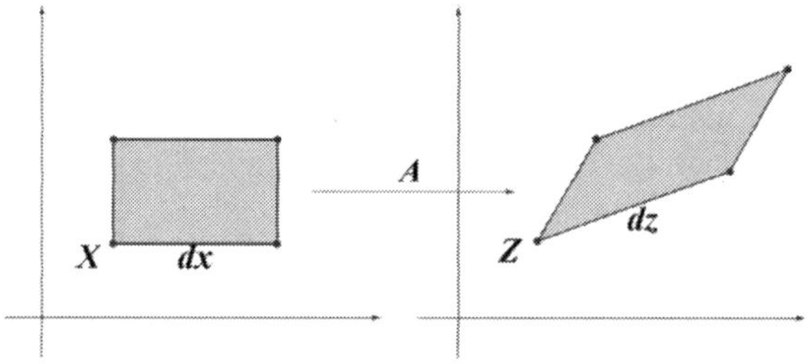


**Figure 1.** Transformation of an Infinitesimal Parallelepiped

Consider the infinitesimal rectangular parallelepiped $\boldsymbol{dx}$ shown in figure 1. Points in this rectangular parallelepiped are mapped to the parallelepiped $\boldsymbol{dz}$. The infinitesimal rectangular parallelepiped and the parallelepiped are equivalent events, so their probabilities must be equal

$$p_X(X)|dX| = p_Z(Z)|dZ|,$$

where $|dX|$, $|dZ|$ are the areas of parallelepipeds. Thus, the joint probability density function for $Z$ is defined as

$$p_Z(Z) = p_X(X)/(|dZ|/|dX|), \tag{3}$$

where random vectors $X$ and $Z$ are related to formula (2).

Equation (3) states that the joint probability density function for the random vector $X$ is the joint probability density function for the random vector $Z$ at the corresponding point, but it is recalculated using the "stretch factor" $|dP|/|dX|$. Next, we will consider the general case $m < n$. For this purpose, we represent model (1) in the form

$$\begin{aligned} z_1 &= f_1(x_1, \dots, x_n), \\ &\vdots \\ z_m &= f_m(x_1, \dots, x_n), \\ z_{m+1} &= f_{m+1}(x_1, \dots, x_n), \\ &\vdots \\ z_n &= f_n(x_1, \dots, x_n). \end{aligned} \tag{4}$$

Suppose that the functions $f_1, \dots, f_m$ are one-to-one continuously differentiable and there is an inverse transformation

$$X = F^{-1}Z_1.$$

The differential of system (4) can be represented as

$$dZ_1 = F'(X)dX,$$

where $F'(X)$ is matrix of first derivatives. Then the joint probability density function $f_{Z_1}(Z_1)$ for a random vector $Z_1$

$$f_{Z_1}(Z_1) = f_X(F^{-1}Z_1)/|F'(F^{-1}Z_1)|.$$

Since, in the general case, the random vector $Z_1 = (z_1, \dots, z_m, \dots, z_n)$ contains fictitious components $z_{m+1}, \dots, z_n$, then to calculate the joint probability density function $f_Z(Z)$ it is necessary to integrate over fictitious variables.

In the next section we will illustrate this statement by the example of solving a system of nonlinear equations with random parameters.

## 3 System of nonlinear equations

Let us investigate the problem of calculating the solution set for a system of nonlinear equations

$$f_i(x, k) = 0, i = 1, \dots, n,$$

where $x \in R^n$ is the vector of solutions, $k \in R^m$ is the vector of parameters. Relatively $k \in R^m$, we will assume that the probability density functions are known. We will consider the case $m \geq n$. Note that the case of strict inequality $m > n$ can be reduced to the case $m = n$ of using the results of probabilistic extensions [8].

Let's consider a special case $m = n$. Differentiating the original system of nonlinear equations, we obtain

$$F'_x(x, k)dx + F'_k(x, k)dk = 0,$$

$$dx = -(F'_x(x, k))^{-1}F'_k(x, k)dk.$$

where $F(x, k) = (f_1(x, k), \dots, f_n(x, k))$.

Thus, knowing the solution $x$ for some parameter values $k$, it is possible to obtain the relationship $dx$ and $dk$.

This fact allows us to calculate the value of the joint probability density function of the solution. Consider an example of a system of nonlinear equations

$$x^2 + y^2 - r^2 = 0,$$

$$xy - c = 0,$$

where $r, c$ are uniform random variables whose probability densities have supports $[1, 1.1]$, $[0.4, 0.5]$.

Let $x_0, y_0$ be the solution of the system for some values $r_0, c_0$, then it is true

$$\begin{pmatrix} 2x_0 & 2y_0 \\ y_0 & x_0 \end{pmatrix} \begin{pmatrix} dx \\ dy \end{pmatrix} = \begin{pmatrix} 2r_0 dr \\ dc \end{pmatrix}.$$

In this case, the rectangle $S_0$ sides $dr, dc$ goes into a quadrangle $S_1$ with sides $dx, dy$. Having solved the system, we get the values $dx, dy$. The probability density of the set of solutions at a point $(x_0, y_0)$ is proportional to the ratio of the areas $|S_0|/|S_1|$:

$$p(x_0, y_0) = p_1(r_0, c_0)|S_0|/|S_1|.$$

Thus, in order to construct a joint density function of random variables , we construct grids in the region of carriers of random variables $r, c$: $\{r_i, i := 0, \dots, m\}$, $\{c_i, i := 0, \dots, m\}$.

Let us solve $(m + 1)^2$ systems of nonlinear equations:

$$x^2 + y^2 - r_i^2 = 0,$$

$$xy - c_j = 0, i, j = 0, \dots, m.$$

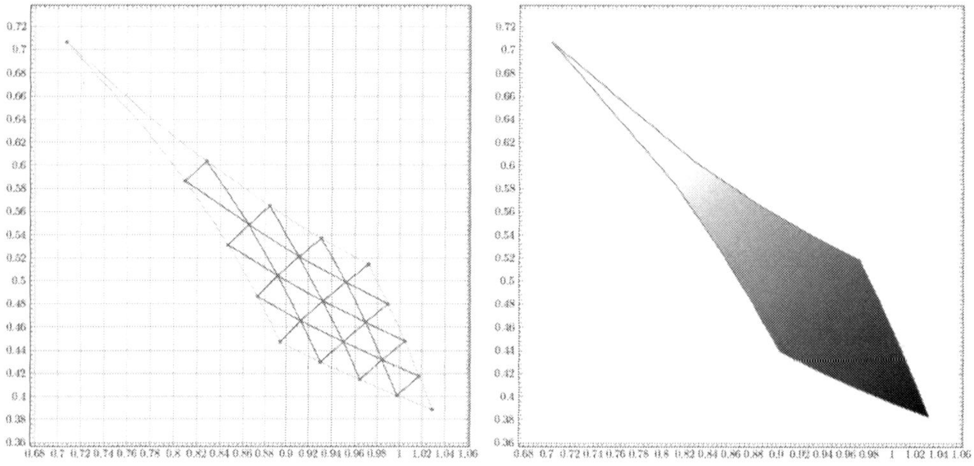

**Figure 2.** Triangulation of the domain and joint density function of the solution set

Figure 2 shows the triangulation of the solution domain for a system of nonlinear equations. The vertices of the triangles are the decision points with the calculated values of the density of the joint probability function of the solution $p(x_i, y_j)$. Thus, using linear interpolation, on each triangle, the value of the joint probability density function can be plotted. Shades of gray represent the joint density function of the solution to a system of nonlinear equations.

Note, to construct the joint density function of the solution, it is required to solve the system of nonlinear equations only 25 times; for the same purpose, the Monte Carlo method will require testing $10^6$. The general case of constructing probability densities can be realized by constructing probabilistic extensions.

## 4 Conclusion

The case is considered when the modeling relationships of input and output variables are presented in the form of systems of equations, both linear and nonlinear, with random uncertainties. The use of CPA and the transformation method makes it possible to find joint probability density functions of the output parameters of remote sensing problems. The method for constructing the joint probability density function of the solution removes restrictions on the requirement of independence of the input parameters, provides an analytical and graphical opportunity to construct the probabilistic region of possible values of the output parameters.

The proposed modified approach provides the results of numerical simulation with the required accuracy of calculations, which is guaranteed by the properties of numerical arithmetic procedures developed within the framework of the CPA.

## References

[1] S. Bartalev, V. Egorov, V. Zharko, E. Lupyan, D. Plotnikov, S. Khvostikov, N. Shabanov, Satellite mapping of the vegetation cover of Russia. Moscow: IKI RAN, (2016)

[2] B. Dobronets, O. Popova. Computational Probabilistic Analysis: models and methods. Krasnoyarsk: Siberian Federal University, (2020)

[3] B. Dobronets, O. Popova, Scientific Computing, Computer Arithmetic, and Validated Numerics. Lecture Notes in Computer Science 9553. Springer International Publishing, Cham, 43–53 (2016)

[4] B. Dobronets, O. Popova, Journal of Siberian Federal University, Engineering and Technologies, 9(7), (2016), 960–971.

[5] B. Dobronets, O. Popova, IOP Conf. Series: Journal of Physics: Conf. Series 1015 (2018) 032028 doi :10.1088/1742-6596/1015/3/032028

[6] B. Dobronets, O. Popova, IOP Conf. Series: Materials Science and Engineering 354 (2018) 012006 doi:10.1088/1757-899X/354/1/012006

[7] B. Dobronets, O. Popova, Journal of Siberian Federal University Mathematics and Physics, (2017) 10(1), pp. 16–21.

[8] B. Dobronets, O. Popova, Tomsk State University Journal of Control and Computer Science 47 (2019) 41–48.

[9] A. Leon-Garcia, Probability, Statistics, and Random Processes for Electrical Engineering. University of Toronto: Pearson Education, Inc. (2008)

# Geological conditions of karst development in the Nizhneirensko-Orda area (Perm Krai, Russia)

*Eugene Erofeev, Valery Kataev*

RF Perm State University (PSNIU), 614990, Perm, Russia

**Abstract. The** article presents the results of research on the use of remote sensing methods in combination with general geological methods to study the patterns of karst development. Practical aspects of expanding the possibilities of using remote sensing methods for engineering and geological surveys in territories of karst development are considered. Analysis and interpretation of the research materials were carried out using spatial analysis and mathematical methods. The priority tasks of the research were to evaluate and expand the possibilities of using remote methods as part of a complex of engineering and geological surveys, to combine remote research methods with general geological methods for solving specific practical problems, to establish general laws of the influence of various natural factors on the intensity and parameters of surface karst forms, their analysis and mathematical justification.

**Introduction.** Taking into account the regulatory requirements of SP 11-1057-П [1] for engineering and geological surveys in karst territories in terms of the integration of Earth remote sensing materials (aerial and cosmophotographs) with general geological research methods, the authors, based on the results of karstological interpretation, space radar surveying the earth's surface and drilling materials, a mathematical assessment of the influence of various natural factors on the parameters of surface karst forms is given.

**Research area.** The Nizhneirensko-Orda area, which occupies the right bank of the lower course of the Iren river up to the longitude of the village of Orda, corresponding to the northern part of the Irensky gypsum and carbonate-gypsum karst region, was chosen as the research area [2]. Administratively, it is the territory of Ordinsky and Kungursky municipal districts of Perm Krai.

Geological and structural conditions – platform, the site is located within the Veslyansky swell, complicating the Bym-Kungur depression. The geological section has been studied based on materials from structural exploration wells to depths of 200–400 m. The geological structure includes deposits of the Ufa ($P_1u$), Kungurian ($P_1kg$), and Artinskian ($P_1ar$) stages, covered loose eluvial-deluvial and alluvial sediments. The nature of the distribution of karst and cover deposits predetermined the conditions for the development of karst. Sulphate-carbonate deposits of the Irenian horizon ($P_1ir$), to a lesser extent carbonate deposits of the Filippovsky horizon ($P_1fl$) and Artinskian stage ($P_1ar$) karst. In the eastern part, the Irene horizon has been destroyed to various depths and transformed into karst breccia (Q-N), represented by cemented anhydrite and dolomite fragments (Fig. 1).

Karst rock in watersheds are exposed or covered with turf, or stratified deposits of the Solikamsk horizon. In large areas in river valleys, karst rock are overlain by alluvial and eluvial deposits. Karst forms are very diverse: funnel, sinkholes, caves, karst ravine, ponors etc. [2].

**Geological exploration.** The territory of the site and adjacent areas is characterized by a high degree investigation. The research was based on materials from drilling, complex hydrogeological surveys, and special works on studying the karst of the Perm Krai territory.

Cartographic constructions are based on the materials of lithological and stratigraphic breakdowns of structural and prospecting (298 pieces) and various hydrogeological (450 pieces) wells that opened the top of the Irene horizon. The section was dissected according to GIS. Information on the surface karst content was obtained from the materials of the CFS interpretation [3-5]. The topographic base was built using space radar imagery (SRTM).

**Geological exploration.** The territory of the site and adjacent areas is characterized by a high degree investigation. The research was based on materials from drilling, complex hydrogeological surveys, and special works on studying the karst of the Perm Krai territory.

Cartographic constructions are based on the materials of lithological and stratigraphic breakdowns of structural and prospecting (298 pieces) and various hydrogeological (450 pieces) wells that opened the top of the Irene horizon. The section was dissected according to GIS. Information on the surface karst content was obtained from the materials of the CFS interpretation [3-5]. The topographic base was built using space radar imagery (SRTM).

The analysis was based on a set of elementary maps based on the materials of generalization and interpretation of the initial data.

 https://doi.org/10.1051/e3sconf/202022302002

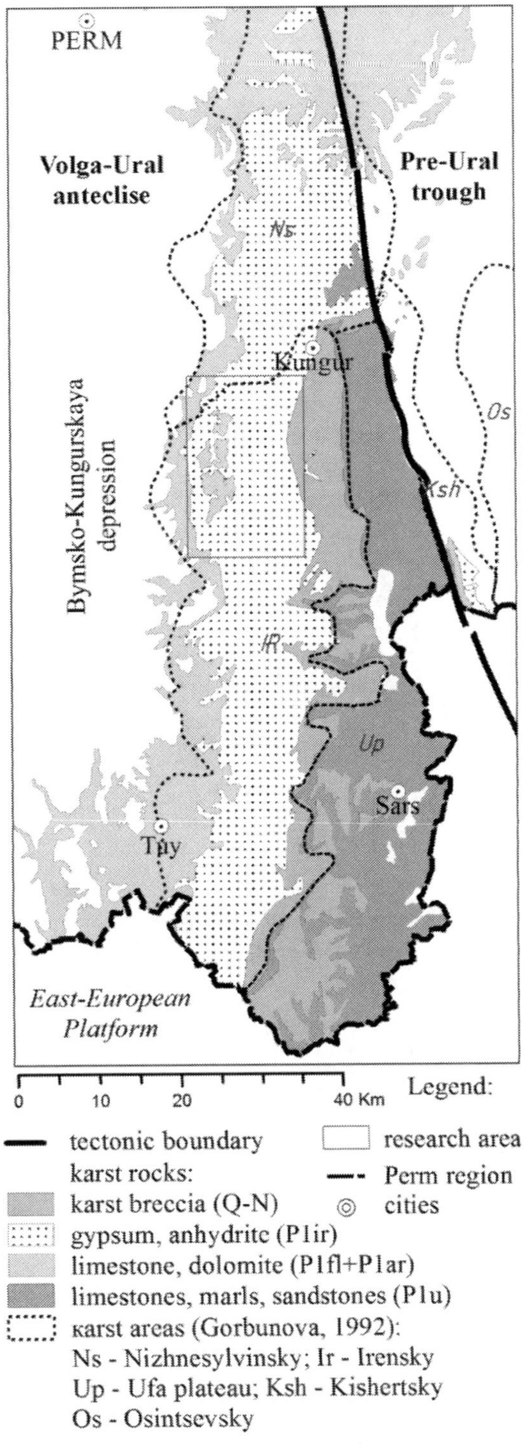


**Fig. 1** Map of karst rocks of the Perm region

The methodological base that determines the approaches to the gradation of the identified factors and the sequence of mathematical calculations was the probabilistic-statistical theory of V.V. Tolmachev [6], some of the methodological provisions of which are discussed in detail in [7]. The processing of research results was carried out using the methods of mathematical methods.

Spatial analysis is based on the following methodological principles: the probability of the formation of disrupted surface forms is controlled by the influence of numerous natural factors, where the role of each individual factor can manifest itself in the form of a certain tendency, which can be expressed quantitatively by a system of coupling coefficients Qi.

Gradation Qi (-1 ... + 1) reflects the probability of occurrence of failures: from –1, the formation of failures with the i-th attribute of the factor under study is impossible, up to +1, on the contrary, the occurrence of failure forms is possible only with the i-th attribute.

The ranking and assessment of the degree of influence of the selected natural factors on the intensity of the formation of karst sinkholes in general can be performed using the concept of entropy from information theory. The informational entropy E of the event (x) is the sum (with a minus sign) of all the products of the relative frequencies of occurrence of the event i, multiplied by their binary logarithms. Comparing the ratio of entropies (Emax / E) of several factors, one can determine the degree of their significance in relation to the influence of factors on the formation of sinkholes (surface karst forms) [6].

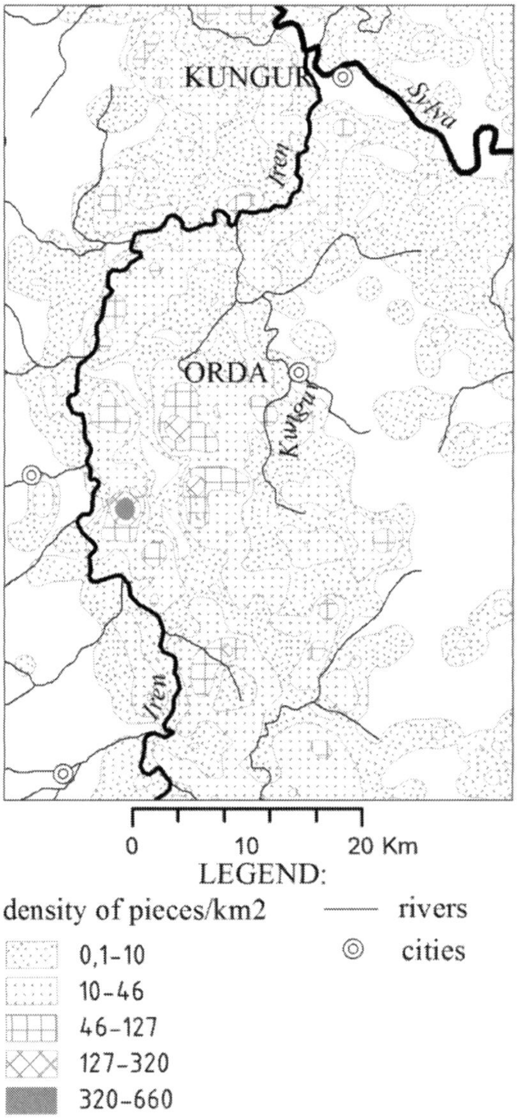


**Fig. 2** Density map of karst forms

**Research result.** Almost 19 thousand forms were identified within the boundaries of the area, the cross-section size was set for 7384 pieces. The density of the karst forms varies from 0.1 to 660 units/km2 (Fig. 2). Spread of values is due to a combination of various geological factors.

The largest number of forms (10469 forms) corresponds to the area of development of the Lunezhka pack, 2039 forms – Tyuiskaya, 2432 – Demidkovskaya, 20 – Elkinsko-Shalashninskaya. A significant number of forms were also recorded within the distribution of the solikamsk cover deposits – 2255 forms. The maximum density of the karst forms corresponds to the boundary between the Irenian and Solikamsk rocks, where the karst rocks are covered only by a thin cover of loose sediments. The most karst-covered are the slopes of river valleys and watersheds of the Iren and Kungur rivers, cut by a gully-girder network (Fig. 2).

This regularity is well traced by the nature of the decrease in the number of karst forms as these forms are removed from the main and smaller rivers. The analysis of remoteness shows that the erosion dissection of the relief (depth of incisions, density of the modern hydro network) is a localizing factor in the development of the karst process. As you move away from the drains, the number of forms decreases sharply (Fig. 3).

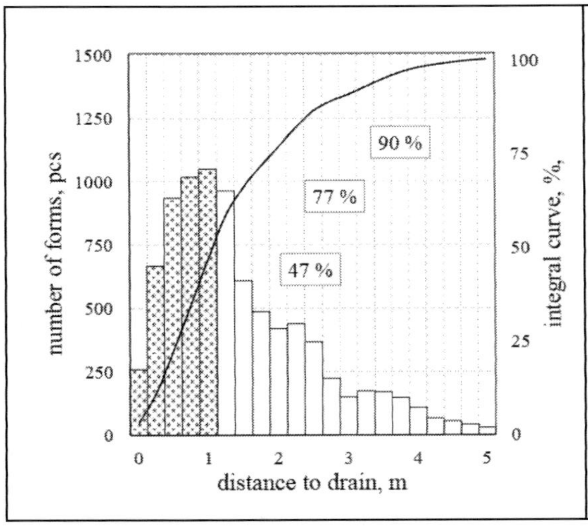


**Fig. 3** Dependence of the number of karst forms on the distance from the river network

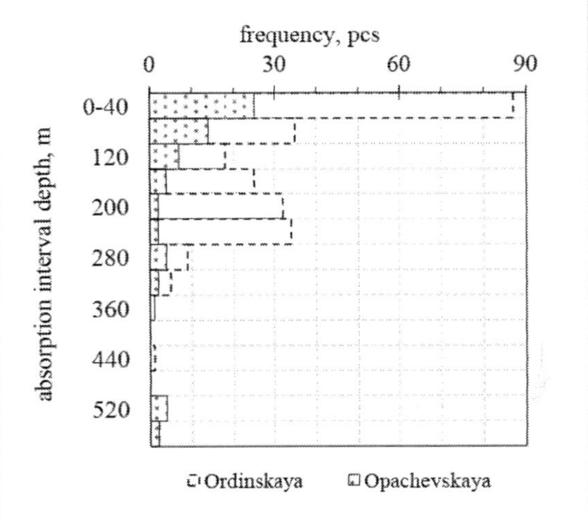


**Fig. 4** Histogram of absorption depth distribution

Almost half (47%) of all identified forms were localized within a narrow kilometer strip bounding the river network, and the remaining 53% were unevenly dispersed within a four-kilometer buffer zone. Also, a change in the diameters of the funnels is noted, for example, with distance from the drain, the average diameter increases from 15.9 m to 23.5 m, and the maximum size from a distance of 2–2.5 km begins to slowly decrease from 90 m to 43 m.

An idea of the nature of the location of underground karst cavities and fractured zones in depth is given by the mud loss diagram of the Orda and Opachevskaya areas (Fig. 3). The maximum number of losses corresponds to the intervals of occurrence of the Irenian horizon of 20–80 m. At the depths corresponding to the Filippov horizon, their number sharply decreases and increases again after driving the top of the Artinskian stage.

The empirical distribution of the transverse dimensions of the karst sinkholes observed at the reference site corresponds to the logarithmically normal law (Fig. 5).

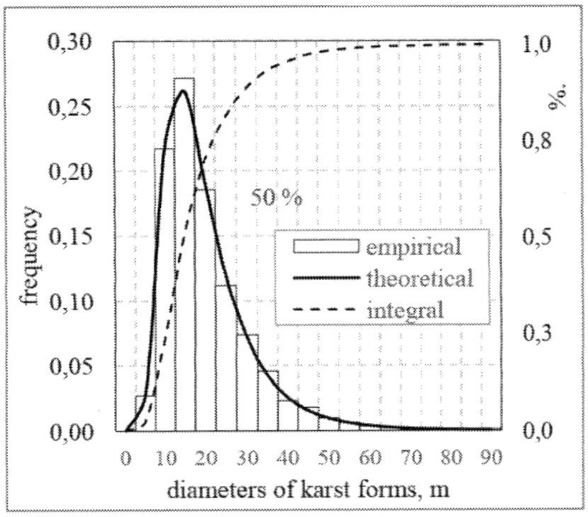


**Fig. 5** Histogram of distribution of diameters of karst forms

Testing the hypothesis is based on the fulfillment of conditions (1-2), with the known assumption that the distribution of the logarithm of the diameters of karst forms obeys the normal law [8]:

$$|A/\delta a| =< 3 \quad (1)$$

$$|E/\delta e| < 3 \quad (2)$$

$$|A/\delta a| = |-0.005/0.029| = 0.16 \quad (3)$$

$$|E/\delta e| = |0.004/0.057| = 0.67 \quad (2)$$

*where, Ek - kurtosis, As - asymmetry, δe and δa standard deviations of skewness and kurtosis estimates, calculated using the built-in Microsoft Office Excel analysis package.*

In the Nizhneirensko-Orda area, craters and sinkholes with a transverse size of 10–15 m prevail, the total number of which is 50% of the total, and the average diameter reaches 17.5 m (Fig. 5)

**Mathematical assessment of the influence of natural factors on the intensity of the formation of karst forms.**

1. The main trends in the development of karst associated with the morphological features of the relief (surface or elevation above the drainage level, slope and energy) are integral indicators of erosional dissection and control the distribution of surface runoff and the depth of drainage of karst waters. The regularity of changes in the surface karst content with an increase in the relief surface is clearly visible. In the interval of heights of 128-190 m, advising the slopes of river valleys, the number of forms increases, but when the height exceeds 190 m, their number gradually decreases. A similar trend can be traced for the slopes of the relief. The growth in the value of the energy potential reflects a steady tendency towards an increase in forms. The role of the slope exposure, which determines the nature and intensity of snow melting in the spring, is also indicative, since the maximum karst content is manifested on the slopes of the western and southern exposures, and the minimum for the eastern (Table 1).

2. Geological factors that reflect the features of the geological structure of the research area can be divided into two groups: the thickness of geological divisions and the depth of their roof.

**As the depth of occurrence of lithological and Conclusions.** The performed analysis of spatial data allowed us to obtain actual quantitative characteristics of the surface karst phenomena, patterns of occurrence and placement of forms in the plan and section, to identify and mathematically justify the main natural and geological conditions for the development of the karst of the Nizhneirensky-Orda area. The prepared materials will form the basis for zoning the territory according to the conditions of karst development and correspond to the pre-project stage of engineering and geological research.

stratigraphic elements increases, the thicknesses lose their role and influence on the surface karst formation. The dominant role belongs mainly to the upper (Tyuiskaya and Demidkovskaya) members of the Irenian horizon, for which a similar tendency is revealed – with increasing thickness, an increase in the number of karst content is observed.

For the uppermost and most destroyed member - Lunezhskaya, the trend is similar, but the strength of the factor is minimal (Table 1).

The second, in its own way, values are indicators of the thickness of the Filippovsky horizon and loose cover deposits. In both cases, an increase in the thickness of deposits is accompanied by a decrease in the number of forms.

3. The indices of the depths of the top of karst deposits have higher coefficients than the geological factors of the first group. With an increase in the depths of the beds of the Irene horizon, their strength and degree of influence decrease. The maximum Emah / E indices belong to the depth factors of the top of the Irenian horizon and the Artinskian stage. With diving below certain marks, the overall karst content decreases (Table 1).

**Table 1. Summary of coefficients Emax / E**

| Factor | $E_{MAX}/E$ |
|---|---|
| Geomorphological factors | |
| Relief surface | **1,0982** |
| Terrain slopes | 1,0118 |
| Slope exposure | 1,0131 |
| Relief energy | 1,0270 |
| Geological factors | |
| Capacities | |
| Irene horizon | 1,0399 |
| Filippovsky horizon | **1,0677** |
| Lunezhskaya pack | 1,0442 |
| Tuy pack | **1,0849** |
| Demidkovskaya pack | **1,0886** |
| Elkinsky pack | 1,0229 |
| Shalashninskaya pack | 1,0493 |
| Covering deposits | **0,0565** |
| Roof depth | |
| Irene horizon | **1,1487** |
| Filippovsky horizon | 1,0858 |
| Lunezhskaya pack | **1,1543** |
| Tuy pack | **1,1469** |

| | |
|---|---|
| Demidkovskaya pack | **1,1215** |
| Elkinsky pack | 1,0705 |
| Šhalashninskaya pack | 1,0227 |
| Artinskiy tier | **1,1762** |

The mathematical tendency for the depth of the Filippovian horizon is identical to the Artinskian stage, but the strength of its influence is much less.

**Mathematical assessment of the nature and degree of influence of natural factors on the transverse dimensions of surface forms.**

The assessment and justification of the choice of natural factors that have an impact on the size of karst forms is based on testing the hypothesis about the homogeneity of the samples of the diameters of the factor signs among themselves, for their belonging to the same general population. If all samples of a factor belong to the same general population, then we can assume that this factor has practically no effect on the size. For verification, the Komogorov-Smirnov criterion was used [8], which showed that all the factors used in the work (Table 1) are capable of influencing the size of karst forms, which in principle is consistent with their distribution law.

**The mathematical assessment of determining the degree of significance of the factors** under consideration by the strength of their influence on the size of the shape was calculated using the coefficient of mutual conjugation of Chuprov (T) [8]. This paper presents the results of the analysis only for the most significant factors in Table. 1, including the thickness of the overburden (Table 2):

**Table 2. Summary table of the values of the contingency coefficient T**

| Factor | T |
|---|---|
| Capacities | |
| Lunezhskaya pack | 0,0482 |
| Covering deposits | 0,0560 |
| Relief surface | 0,0695 |
| Roof depth | |
| Irene horizon | 0,0653 |
| Tuy pack | 0,0596 |
| Demidkovskaya pack | 0,0594 |
| Artinskiy tier | 0,0570 |

**Conclusions.** The performed analysis of spatial data allowed us to obtain actual quantitative characteristics of the surface karst phenomena, patterns of occurrence and placement of forms in the plan and section, to identify and mathematically justify the main natural and geological conditions for the development of the karst of the Nizhneirensky-Orda area. The prepared materials will form the basis for zoning the territory according to the conditions of karst development and correspond to the pre-project stage of engineering and geological research.

**List of references**

1. SP 11-105-97 Engineering and geological surveys for construction. Part I. General rules of work. Russia (1998).
2. K.A. Gorbunova, V.N. Andreychuk, V.P Kostarev, N.G. Maksimovich. Karst and caves of the Perm region. Publishing house Perm. University, Russia (1992).
3. E.A. Erofeev, V.N. Kataev. Identification of surface karst forms based on satellite images. Geology and minerals of the Western Urals: collection of articles of the 37th All-Russian scientific and practical conference. May 23-24 2017, Perm, Perm University, Russia (2017).
4. E.A. Erofeev, V.N. Kataev. Estimative karst interpretation of satellite images of the south – south-eastern districts of Perm region. International Symposium KARST 2018 – Expect the Unexpected. Trebinje, Bosnia and Herzegovina (2018).
5. E.A. Erofeev, V.N. Kataev. Methodical approach to carrying out an estimated karstological interpretation of the territory of karst regions of the Perm Territory. Regional problems of remote sensing of the Earth. Materials of the VI International Scientific Conference. Krasnoyarsk, September 10-13, 2019. Russia (2019).
6. V.V. Tolmachev, G.M. Troitskiy, V.P. Khomenko. Engineering and construction development of karst territories. Stroyizdat, Russia (1986).
7. E.A. Erofeev, V.N. Kataev. Application of probabilistic and statistical methods for assessing karst hazard under conditions of anthropogenic impact on karst territories. Engineering geology. December 4/2010. from 34-46. Russia (2010).
8. I.P. Sharapov. Application of mathematical methods in geology. Publishing house "Nedra", Russia (1971).

# Spatial routing network building for pedestrian panoramic survey points when monitoring urban areas

*Ignat N.* Penshin[1,], *Larisa G.* Evstratova[1]

[1]State University of Land Use Planning, 105064, 15 Kazakova str., Moscow, Russian Federation

**Abstract.** The relevance of the development of new solutions for building a spatial routing network for points of pedestrian panoramic shooting when monitoring urban infrastructure is considered. A computationally efficient automated algorithm has been developed and implemented in software, which provides the ability to move the user in the forward and backward direction between any pair of incident points of a panoramic survey using a custom geoportal. The proposed approach takes into account changes in the urban environment (transport networks, capital structures), thus, they are not an obstacle to the operation of the created spatial routing network. It remains possible to rebuild the network at any required scale (shooting tracks, groups of tracks, urban areas, administrative districts, the whole city).

## 1 Introduction

Today, there are various web services for viewing street panoramas, in which images obtained by means of automobile panoramic shooting are available. However, this does not provide a complete overview of urban areas [1]. This disadvantage can be eliminated by pedestrian panoramic photography, since it provides the maximum possible territorial coverage due to the high mobility of the operator performing the shooting, i.e. positioning independent of the road network. In this regard, the automation of spatial routing within courtyard areas for points of pedestrian panoramic survey becomes relevant.

The effectiveness of urban resource management depends on the complete and timely receipt of information about the objects of urban infrastructure, which is achieved through monitoring [2]. A georeferenced and spatial-oriented panoramic image allows to get up-to-date information about the city space. Interpretation of the obtained data array during panoramic shooting within the framework of geographic information system allows to conduct an inventory and subsequent analysis of the compliance of certain territories using with the intended purpose [3, 4].

[1] Corresponding author: ignat.penshin51@gmail.com

## 2 Navigation network construction algorithm for pedestrian panoramic survcy points

The functional block of an intelligent GIS agent contains an algorithm that generates a set of rational solutions when creating spatial routing inside courtyard areas without using information about the road network [5].

The algorithm takes initial data as input and returns a set of links between point objects. Figure 1 shows a block diagram of the sequence of actions of the algorithm for spatial routing network construction. Next, we will consider the main factors affecting the complexity of the algorithm.

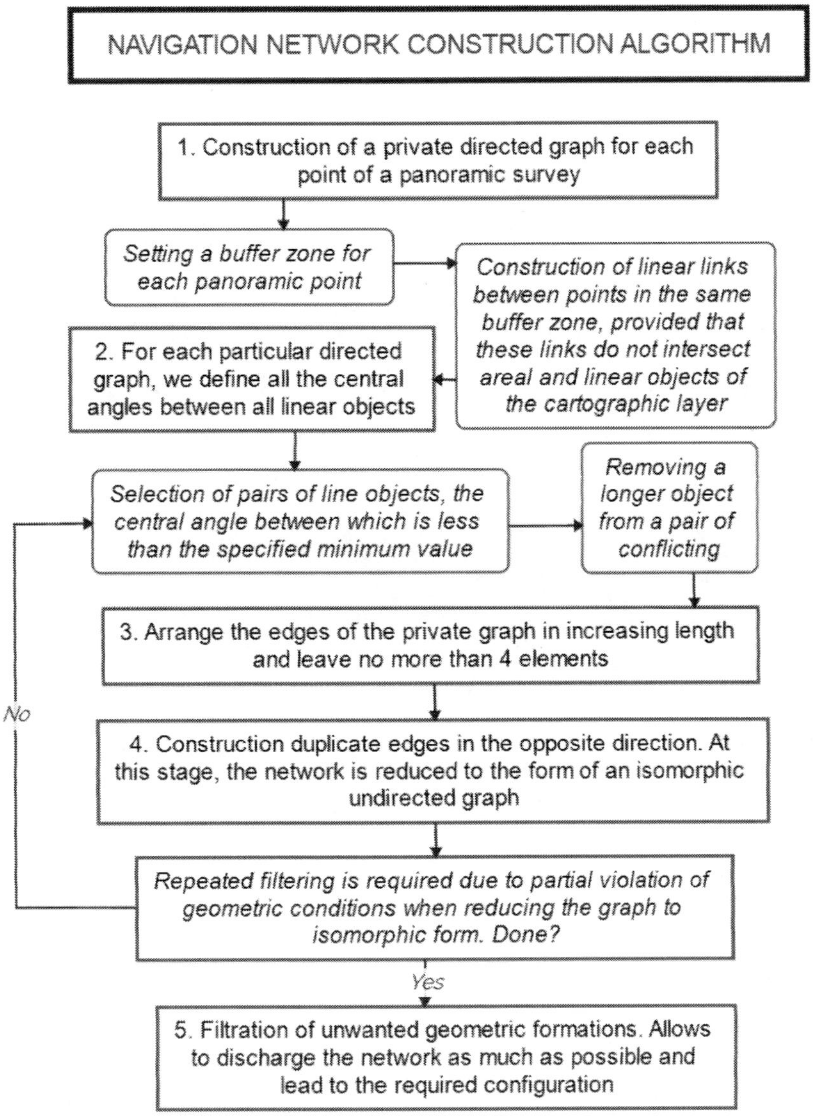


**Fig. 1.** Block diagram of the spatial routing network construction algorithm

 https://doi.org/10.1051/e3sconf/202022302003

## 3 Analysis of the decision rules complexity

The greatest contribution to the complexity of the algorithm is made by the execution of 1 decision rule. The choice of the optimal parameters for its execution significantly reduces the amount of computation. Let us consider in more detail the order of execution of 1 decision rule, and also define the input sets of objects:

During the execution of decision rule 1, a buffer zone with a radius equal to the variable of maximum distance between points (buff_radius = dist_max) is created for each panoramic point (Panoramas table). Neighborhood points (Panoramas table), intersected by the buffer zone of this point, are used to create an array of linear objects with the beginning at this point - thus we get a private directed graph for this point [5]. Next, line objects crossing linear (Highways table) or area objects (Buildings table) that are in the buffer of this point are deleted. The described sequence of decision rule 1 actions is performed for each point of the Panoramas table.

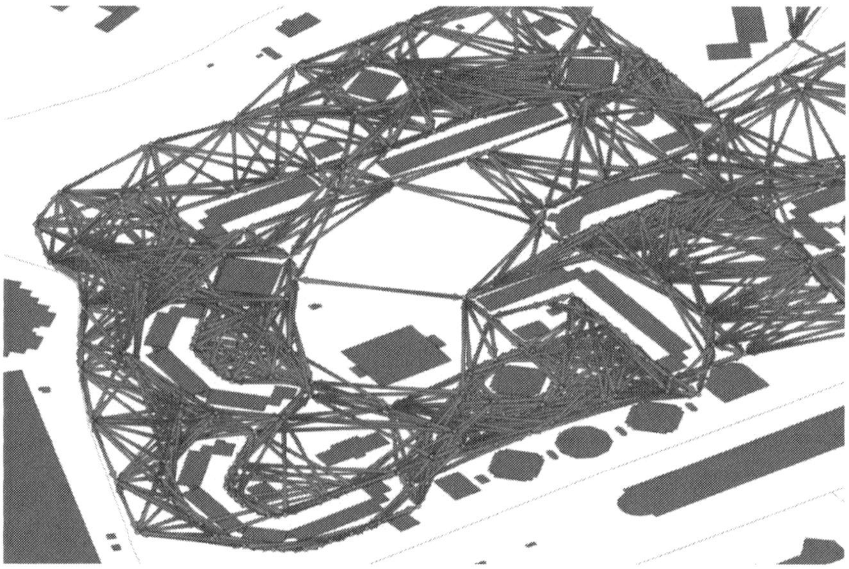

**Fig. 2.** Execution of the 1 decision rule of the spatial routing network construction algorithm

Let us define the set of Panoramas photographing points as V, the set of areal objects Buildings - B, the set of line objects Highways - H, the set of connecting edges Edges - E. If an object of any of the sets V, B, H is in the intersection with the buffer of the given initial point $v_j$, then this object is considered when forming a private graph for this initial point and falls into the subset $\Gamma_{v_j}$. We will use the mathematical apparatus of graph theory and define subsets of a given initial point $v_j$ (1-3):

$$\Gamma_{v_j}(u) = \{\forall u_i \in u \in V \mid max\ s(u_i, v_j) \leq dist_{max}\} \quad (1)$$

$$\Gamma_{v_j}(b) = \{\forall b_i \in b \in B \mid max\ s(b_i, v_j) \leq dist_{max}\} \quad (2)$$

$$\Gamma_{v_j}(h) = \{\forall h_i \in h \in H \mid max\ s(h_i, v_j) \leq dist_{max}\} \quad (3)$$

To determine the above subsets, it is necessary to perform a sequential enumeration of all elements of the sets V, B, H relative to each photographing point as the initial one. The number of elements of this set is directly proportional to the number of operations performed and is determined by its capacity. Then the complexity of performing the first action $M_{1.1}$ to determine the subsets of each point is determined by the formula (4):

$$M_{1.1} = (|V| + |B| + |H|) * |V| \tag{4}$$

After defining the subsets $\Gamma_{v_j}(u)$, $\Gamma_{v_j}(b)$, $\Gamma_{v_j}(h)$, the subset of connecting edges $E_j$, implemented as an array of linear objects, is determined. This subset is described by formula (5):

$$E_j = \left\{\forall u_i \in \Gamma_{v_j}(u) \Lsh (v_j, u_i)\right\} \tag{5}$$

Satisfaction of the condition of absence of intersection of connecting edges $E_j$ and objects of subsets $\Gamma_{v_j}(b)$, $\Gamma_{v_j}(h)$ is defined by the subset of edges $E_j^{'} \in E_j$. The subset $E_j^{'}$ corresponds to a logic judgment of the form: the set of edges $E_j$ that do not belong to the subset of areal objects $\Gamma_{v_j}(b)$ AND the set of edges $E_j$ that do not belong to the subset of linear objects $\Gamma_{v_j}(b)$. The logical notation $E_j^{'}$ will take the form (6):

$$E_j^{'} = E_j \backslash (\forall b_i \in \Gamma_{v_j}(b)) \wedge E_j \backslash (\forall h_i \in \Gamma_{v_j}(h)) \tag{6}$$

To implement the condition of the absence of intersections and the formation of the subset $E_j^{'}$, it is necessary to perform a sequential enumeration of the area objects Buildings from the subset $\Gamma_{v_j}(b)$ and linear objects Highways from the subset $\Gamma_{v_j}(h)$ for connecting edges from the set $E_j$. The complexity of the second action $M_{1.2}$ is determined by the formula (7):

$$M_{1.2} = \sum_{j=1}^{|E_j|} \left(\left|\Gamma_{v_j}(b)\right| + \left|\Gamma_{v_j}(h)\right|\right) \tag{7}$$

The complexity of the sequential performance of actions in the first decision rule can be estimated by the formula (8):

$$M_1 = \left( |V| + |B| + |H| + \sum_{j=1}^{|E_j|} \left(\left|\Gamma_{v_j}(b)\right| + \left|\Gamma_{v_j}(h)\right|\right) \right) * |V| \tag{8}$$

From formula (8) it follows that the number of operations performed is directly proportional to the size of the processed area, namely, the number of survey points, construction objects and roads included in the area. If we assume that the difference in resource intensity and execution time of computational operations is negligible, then we can compare the number of operations for processing input data arrays that differ in size. Further actions will make it possible to determine the optimal size of a unit processing area based on an a priori estimate of the complexity of the decision rule 1 execution by an intelligent agent [6, 7].

According to the averaged performance indicators of the field group of operators of pedestrian panoramic survey, as well as separate averaged representations of geodata, we will use the following constant values of the quantities involved in determining the complexity of the first decision rule (Table 1).

**Table 1.** Constants in determining the complexity of decision rule 1

| Value | Quantity |
|---|---|
| $\overline{\lvert E_j \rvert}$ - the average number of Panoramas vertices falling into the buffer zone of this initial vertex | $\sim 8$ |
| $\overline{\lvert \Gamma_{v_j}(b) \rvert}$ – the average number of areal Buildings objects that fall into the buffer zone of this initial vertex | $\sim 3$ |
| $\overline{\lvert \Gamma_{v_j}(h) \rvert}$ - the average number of Highways line features that fall within the buffer zone of this initial vertex | $\sim 2$ |

Let's assume that, on average, the territory of one block contains: 10 areal Buildings, 4 adjacent highways, 200 Panoramas panoramic points. The calculation of the complexity of fulfilling 1 decision rule according to formula (8) for territories that are multiples of the size of one quarter is presented in Table 2.

**Table 2.** Decision rule complexity 1 for input Datasets

| Sets | Dataset 1 | Dataset 2 | Dataset 3 | Dataset 4 | Dataset 5 | Dataset 6 |
|---|---|---|---|---|---|---|
| $\lvert V \rvert$ | 5000 | 2500 | 1000 | 500 | 250 | 100 |
| $\lvert B \rvert$ | 250 | 125 | 50 | 25 | 12 | 5 |
| $\lvert H \rvert$ | 100 | 50 | 20 | 10 | 5 | 2 |
| **Complexity (number of operations)** | | | | | | |
| $M_1$ | 26 950 000 | 6 787 500 | 1 110 000 | 287 500 | 76 750 | 14 700 |

According to the values of $M_1$ for different sets of input data, the most optimal in terms of time and productivity is the yard processing of the territory when the decision rule 1 is fulfilled. The optimal sets correspond to Set 5 and Set 6 of Table 2.

## Conclusion

Experiments on the performance of the proposed spatial routing algorithm for panoramic survey points show that the proposed approach generates a set of rational solutions when creating a spatial routing network inside courtyard territories without using information about the road network and the most optimal in terms of time and productivity is quarterly processing of the territory. As a result, the geoportal, which aggregates and provides the user with georeferenced and spatial-oriented panoramic images, makes it possible to solve the following applied problems: analytics in the field of urban planning without the need for physical presence in the study area; inventory of objects of engineering, transport, social, industrial, residential infrastructure, development of the recreational potential of territories, etc.

## References

1. Nam, T., & Pardo, Conceptualizing smart city with dimensions of technology, people, and institutions. In Proceedings of the 12th annual international digital government research conference: digital government innovation in challenging times, 282 (2011)
2. Signore R., Stegman M. O., Creamer J. The ODBC solution: Open database connectivity in distributed environments. – McGraw-Hill, Inc., 1995
3. Fotheringham, S., Rogerson, P., Spatial analysis and GIS (2014)

https://doi.org/10.1051/e3sconf/202022302003

4. Hand, D. J., Data Mining, Encyclopedia of Environmetrics, **2** (2006)
5. Penshin I. N., Evstratova L. G. Navigation network construction method for survey points using cartographic urban environment information //E3S Web of Conferences, . **149**, 02001, (2020)
6. Xiaoping D. P. Y. C. R., Jiliang G. Algorithm of Spatial Select Query in GIS Based on the Improved Quadtree with the Case of ESRI SHAPE File [J] //Computer Engineering and Applications, **13** (2003)
7. Piórkowski A. et al. Mysql spatial and postgis–implementations of spatial data standards //Electronic journal of Polish agricultural universities, **14**, 1-8 (2011).

# Remote spectral methods for detecting stress coniferous

*Leonid* Katkovsky [1*], *Boris* Beliaev[1], *Volha* Siliuk[1], *Mikhail* Beliaev[2], *Erik* Sarmin[2], and *Yurii* Davidovich[1]

[1]A.N. Sevchenko Institute of Applied Physical Problems of Belarusian State University, Minsk, 220045, the Republic of Belarus
[2]S.P. Korolev Rocket and Space Public Corporation Energia, Korolev, 141070, Russia

**Abstract.** The article presents investigation of the possibility of drying coniferous forest areas detecting by multispectral satellite data in the visible and NIR spectral range with low spatial resolution, obtained by the imaging systems of three satellites - the Belarusian spacecraft (BS), Landsat 8 and Sentinel 2. A forest area in the south of Belarus was considered as a test site. High-resolution multispectral airborne data and, in part, ground measurements were used as reference ground data by which training samples were formed. Most of the known classical methods of supervised classification have been tested, the maximum likelihood method turned out to be the best for this task. In order to improve the accuracy of identifying the drying areas of coniferous forests on multispectral images, parametric transformations of images in the spectral space are proposed, which should lead to an increase in initial small spectral differences. The methodological issues of assessing the accuracy of the satellite images classification are considered using the result of the classification of airborne image with high spatial resolution as a ground truth image. The assessment of the classification accuracy, both visually and using the obtained confusion matrices, allows us to conclude that the images of the BS, Landsat 8 and Sentinel 2 can be used to detect drying area of coniferous forests as well as the expediency of carrying out the proposed transformations of the original images.

## 1 Introduction

Sharp-dentated Bark Beetle (*Ips acuminatus*) has been recognized as one of the most dangerous pests in the EU countries, although it was previously considered an inactive species. Since 2010, a new pathological phenomenon has not been noted earlier on the territory of Belarus has been called "bark beetle drying of the pine". Extreme weather in recent years (decreased precipitation, lower water tables, drought and hurricane winds) which weakens trees is the main reason of pest outbreaks. Identification at early stages of damage to coniferous forests by various types of pests (bark beetle in the central and eastern part of Europe and silkworm in Siberian forests, other insects) is an urgent task of forestry.

* Corresponding author: katkovskyl@gmail.com

Coniferous forests monitoring using satellite remote sensing [1, 2] is preferable due to the wide coverage and the efficiency of obtaining up-to-date information [3]. There are some multispectral satellite sensors with a spatial resolution of 10–30 m, the data of which are in the public domain and usually cover large territories [4] than expensive images of very high (1–10 m) and ultra-high (less than 1 m) resolutions. The interest of this study is finding out possibility of using such data for detecting of stress coniferous forest area.

The paper describes the methods and the results of the detection of drying forest area on satellite multispectral images, using reference samples (regions of interest in the images), which were visually detected on airborne very high spatial resolution images and then confirmed by ground measurements. The following remote sensing data were used: an airborne image obtained by a 4-band ADS-100 Leica camera from a height of several hundred meters (the spatial resolution during processing was roughened to 1 m); a 4-band multispectral sensor image of the Belarusian spacecraft (BS) (spatial resolution 10.5 m); the Landsat 8 OLI (30 m) and Sentinel 2 (10 m) satellite data (for both were taken 4 band lying in the visible and near-infrared spectral regions). All data were georeferenced. Airborne image, in contrast to the used satellite images, was in digital numbers (without radiometric correction, due to the lack of appropriate camera calibration). At first, satellite data were used for classification without atmospheric correction (in top-of-atmosphere radiance units), then - the same images after atmospheric correction (in bottom-of-atmosphere reflectance units).

In the first case (radiance images), in order to improve visualization and the possibility of making comparisons with the airborne image, pan-sharpening of the BS and Landsat 8 OLI (in the ENVI software package) was previously performed. After performing the supervised classification of the initial images, transformations of images in the spectral space were performed to enhance the spectral differences of objects (Section 2.1), which were also subjected to classification.

## 2 Methods for the classification and detection of stress forest area on satellite images

In the first step, small test region of forest contained stress area were chosen on satellite images (the test region was strictly coincided in all four images). The Landsat 8 image corresponding to the selected area contained only 16 x 11 pixels. In the second step, the selection of regions of interest (ROI) was made, which were used as training samples for each of the class. The classes were identified on visual interpretation of a high-resolution airborne image, confirmed by ground measurements, which was considered as a reference.

When comparing the classification results of images that differ greatly in spatial resolution (airborne and satellite), the problem of methodologically correct selection of ROI and the number of classes arises. For example, small-area "clean" ROIs in an airborne image may occupy only a small fraction of a pixel in a low-resolution satellite image, which is thus a mixture of different classes. At the same time, the choice of the same classes and their number is often problematic due to the difficulty of identifying some objects in low spatial resolution images. In particular, it is difficult to select the "tree shadow" class in a satellite image with a dense forest. One way to avoid this problem was to independently select the ROI on each image, as shown in Figure 1, where the ROIs are shown for the drying class in the airborne image as a background, and the composition of the classes was different in remote sensing images, but the drying class was present everywhere.

**Fig. 1.** ROI "stress" class, selected on airborne image (red), BS satellite image (orange), Sentinel 2 satellite image (blue) and Landsat 8 (yellow). The airborne image is shown as a background.

The next section 2.1 describes the proposed transformations that were applied to the original images.

## 2.1 Spectral image transformations

If we consider a multispectral image in N-dimensional space with axes representing image bands, then it is represented in it as an N-dimensional "cloud" of points.

The greater the difference in brightness in the bands, the further away the points of the "cloud" are from the bisector of the N-facet angle formed by the coordinate planes. The transformation of a set of points, leading to their distance from the bisector, should lead, in our opinion, to an increase in the differences in the image bands, a decrease in their correlation, and an improvement in separability. The new N-bands image is calculated as:

$$\overline{I}_{k,ij} = T_{ij} I_{k,ij} + (1/N)(1 - T_{ij})\left(\sum_{n=1}^{N} I_{n,ij}\right) \quad (1)$$

Here $I_{n,ij}$, $\overline{I}_{k,ij}$ – original and transformed images respectively, *k*, *n* – bands numbers, *(i, j)* – pixel number (row and column). The image parameter sets the degree of amplification of the initial interband differences of the image and depends on the distance $d_{ij}$ of the current pixel *(i, j)* to the bisector of the N-facet angle:

$$d_{ij} = \sqrt{\frac{1}{N} \sum_{n>m=1}^{N} (I_{m,ij} - I_{n,ij})^2} \quad (2)$$

Two options are proposed for *T*. In the first option $T_{ij}$ is inversely proportional to $d_{ij}$:

$$T_{ij}(d_{ij}) = (T_{max} d_{max}) / \{d_{max} + d_{ij}[(T_{max}/T_{min}) - 1]\} \quad (3)$$

Here $d_{max}$ – maximum value $d_{ij}$, $T_{max}$, $T_{min}$ – set conversion parameters, respectively, maximum and minimum values of $T_{ij}$. It is easy to show that the deviation of the point of the new multispectral image from the bisector is $\overline{d}_{ij} = T_{ij} d_i$. From formula (3) it follows that $Tij(d_{max}) = Tmin$ and $T_{ij}(0) = T_{max}$ (if there are points lying on the bisector), i.e. points lying closer to the bisector move away from it as much as possible, and those lying further away - minimum [5].

As a second option, it is proposed to use a directly proportional relationship $T_{ij}(d_{ij})$ from $d_{ij}$ of the following kind:

$$T_{ij}(d_{ij}) = T_{min}\left[1 + (d_{ij}/d_{max})(T_{max}/T_{min} - 1)\right] \quad (4)$$

Wherein $T_{ij}(d_{max}) = T_{max}$ and $T_{ij}(0) = T_{min}$, i. e. points farther from the bisector are removed farther than closer ones. With regard to a multispectral image, relatively small differences in the bands (points lying near the bisector) do not make sense to significantly enhance, since these differences will usually be caused by measurement errors (noise). At the same time, noticeable differences, even more enhanced, should give better discrimination of objects, based precisely on the values of the brightness in individual bands. The values of the transformation parameters (3), (4), $T_{max}$ and $T_{min}$ are selected empirically by trial calculations. Image transformation and classification were performed in the ENVI program.

## 2.2 Classification results

When classifying, on the image covers a small area of forest four classes were set: stress, healthy coniferous forest, deciduous forest and shadows. The degree of separability of the specified classes was estimated by the selected ROI using two parameters: the Jeffreys-Matusit separability measure and the transformed divergence, the values of which (close to 2) showed very good separability every time. However, a certain methodological problem arises when assessing the classification accuracy of satellite images. If we evaluate it in relation to the selected ROI (ground truth ROI), then it was high in all cases (more than 90%) and is not presented here.

Since the classification accuracy of the original airborne image of the ADS camera was very close to 100% for all classes (according to ROI), it was logical to use the classified ADS image as the ground truth image. For this purpose, in order to compare the classification results, it was assumed that all ROIs should be the same on all images (of different resolutions), while their position and size are determined by geographic coordinates.

Thus, the results of the classification from the airborne images were used as a reference. To construct a confusion matrix for all satellite images from an ADS airborne image as a ground truth image, it was necessary to convert the satellite images to ADS resolution (1 m) and the same number of pixels (Layer stacking operation in ENVI). Finally, it is advisable to group neighboring pixels (clump classes) in a high-resolution classified image (in order to bring the resolutions of the compared images closer together), which often improves the classification accuracy, which is given by the reference image error matrix (ADS).

Examples of the results of image classification are presented in Table 1, where the second column shows the original images, and the images of the BS, Sentinel 2, Landsat 8 OLI are resampled to the resolution of the ADS aerial image. In the next column - the results of classification by the maximum likelihood method - the "drying" class is overlayed on the original images, and before classification one of the transformation options in the spectral space was carried out (formulas (1) - (4)). The last column of the table shows the classification accuracy: general and by the drying class to classify the original image and the best version of the converted image, for ADS the accuracy is given in relation to the original ROI, for satellite images - according to the error matrix, where the ground-based true data is the ADS classification.

At the second stage of the classification, the same images, but in reflectance units (after atmospheric correction, except for ADS) were taken. But the area of forest was much larger (the Landsat image already contained 112x52 pixels), the scene contained a greater number of classes (non-forest objects in Fig. 2 are masked in black)), seven classes were already defined for the forest instead of four.

Figure 3 shows the classification results for ADS and BS (overlaying the drying class on the converted original images.

**Table 1.** Results of determining forest stress from transformed multispectral images by the maximum likelihood method.

| **Imaging system** | **Original images, satellite - after stretching to ADS resolution** | **ADS and satellite classification results converted by one of the options** | **Accuracy of classification, total / according to the drying class according to the original / according to the transformed,%** |
|---|---|---|---|
| ADS | 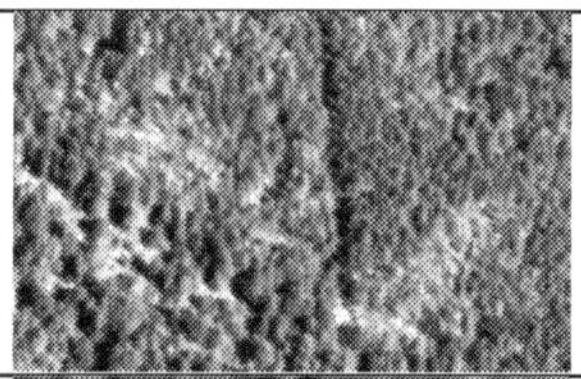 | 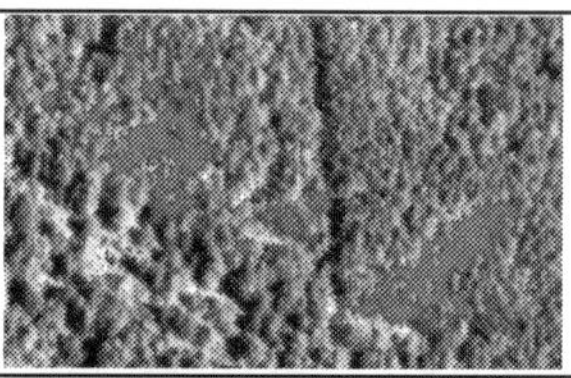 | 99,6 / 100 |
| BS | 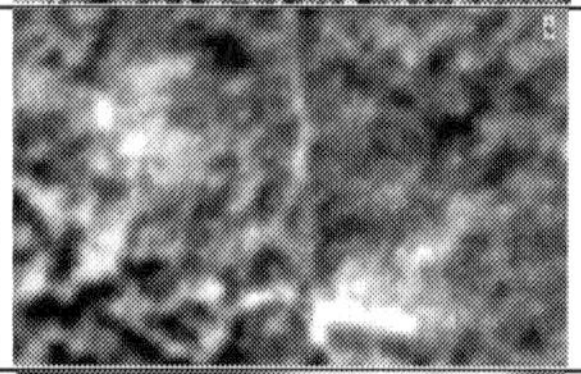 | 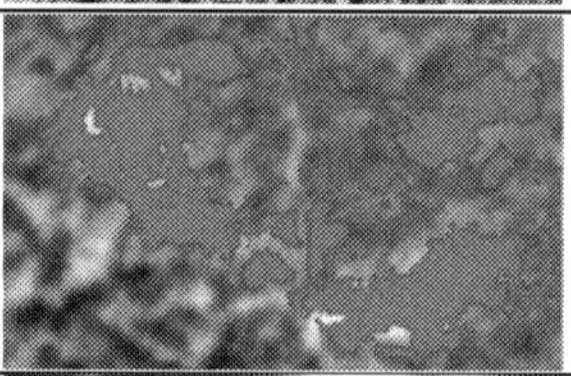 | 51,6 / 57,8 / 60,9<br>After clump classes by 7x7 pixels: 96,8/85,6 |
| Sentinel 2 | 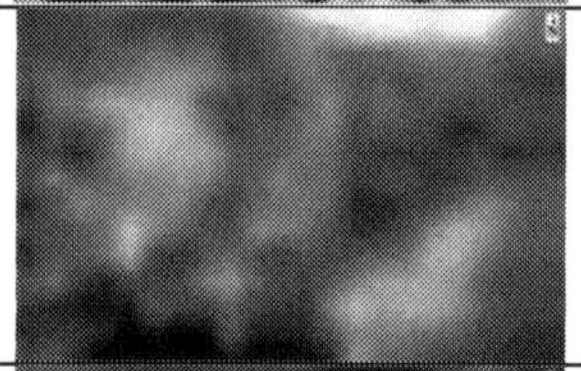 | 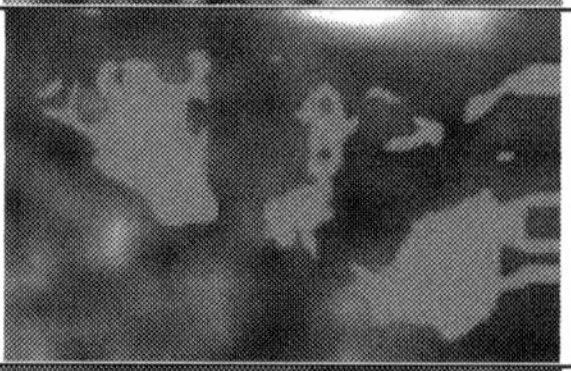 | 52,5 / 56,5 / 57,5 |
| Landsat 8 OLI | 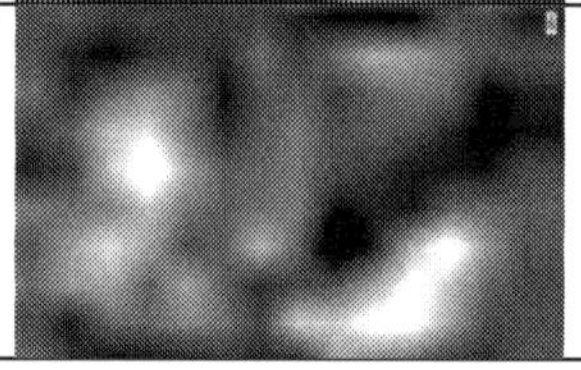 | 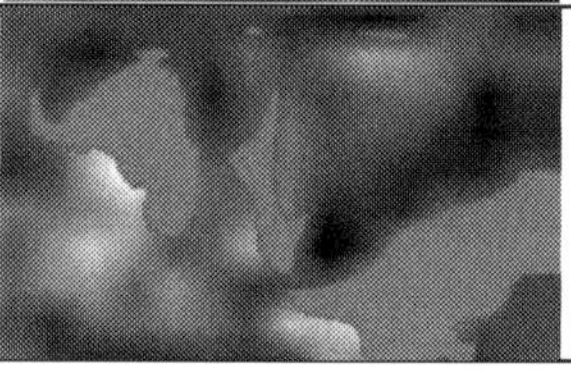 | 55,7 / 56,8 / 57, 2 |

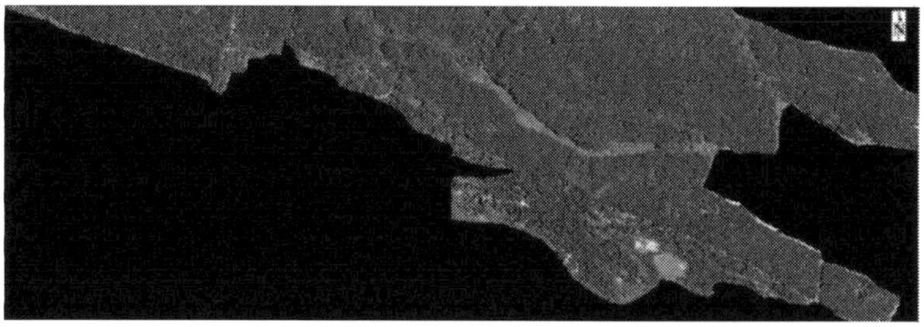

**Fig. 2.** Overlay ROI "drying" highlighted in airborne image (red)

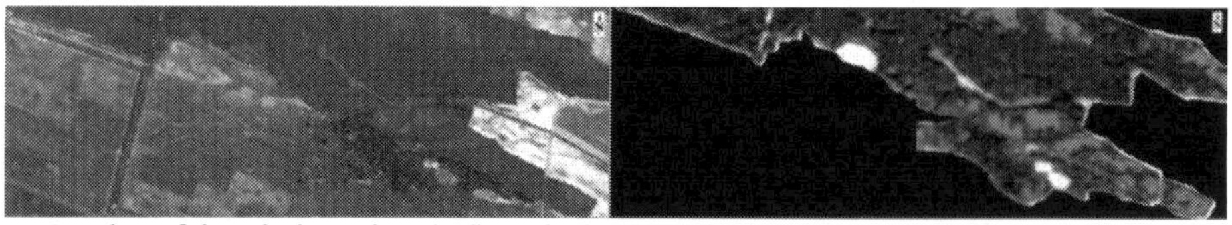

**Fig. 3**. Overlay of the "drying" class (red) on the images of ADS (left) and BS (right)

The classification accuracy of the ADS image relative to the target ROI was 99.8% (overall) and 81.9% for the "drying" class. For satellite imagery, the accuracy relative to the ADS classification of the original and converted imagery is shown in Table 2.

**Table 2.** Accuracy of satellite imagery classification (original and transformed) relative to ADS classification

| **Imaging system** | **Classification accuracy, total/drying, % (original images)** | **Accuracy of classification, total / drying, % (converted images)** |
|---|---|---|
| BS | 94,1 /17,1 | 94,5 / 24,6 |
| Sentinel 2 | 95,3 / 14,5 | 97,4 / 16,5 |
| Landsat 8 OLI | 94,3 / 24,5 | 94,4 / 23,3 |

## 3 Conclusion

The obtained classification results allow us to conclude that the images of the BS, Landsat 8 and Sentinel 2 can be used for reliable detection of drying several tens of meters areas of coniferous forests, while it is advisable to carry out the proposed transformations of images in the spectral space, which enhance their discrimination. The benefits of using images after atmospheric correction were not clearly identified, but this may be due to different image sizes before and after correction, and to resampling (strong smoothing).

## References

1. H. Abdullah, R. Darvishzadeh, A.K. Skidmore, M. Heurich, Remote Sens. 11 (2019).
2. V.I. Kharuk, S.T. Im, M.L. Dvinskaya, Russian Journal of Ecology, 3 (2016).
3. I.A. Saigin, S.A. Bartalev, F.V. Stytsenko, A method for detecting long-term drying out of dark coniferous forests in Russia based on satellite data, in Proceedings of the 17th All-Russian Open Conference "Current Problems of Remote Sensing of The Earth From Space", Space Research Institute, 11-15 November 2019, Moscow, Russia (2019).
4. Yu.S. Davidovich, L.V. Katkovsky, Recognition of tree vegetation stress in aerospace imagery, in Proceedings of the GIS technologies in geosciences, BSU, 13 November 2019, Minsk, Belarus (2019).
5. B.I. Belyaev, L.V. Katkovskiy, Optical remote sensing, BSU, (2006).

# About the geographic grid of the three-dimensional hypersphere

*Pavel* Kim[1,*]

[1]Federal State Budgetary Institution of Science Institute of Computational Mathematics and Mathematical Geophysics, Siberian Branch of the Russian Academy of Sciences, 6, Ac. Lavrentieva ave., Novosibirsk, 630090, Russia

**Abstract.** According to the Poincaré conjecture (1904) proved by Grigory Perelman (2002-2003) that any simply connected compact three-dimensional manifold without edges is homeomorphic to a three-dimensional hypersphere [1], to solve the problems of visualizing four-dimensional objects in three-dimensional space [2], it is proposed to choose a suitable manifold, in in this case, a ball, establishing a homeomorphism between objects located in different spaces by technological means of cartography. As a result of this work, it seems possible to build a dynamic video of the population distribution process on a map of the globe, which provides informational four-dimensional data flow, following the ideas embodied in 4D Anatomy [3]. The proposed technology opens up new ways of visualizing four-dimensional space This work was performed within the framework of the state assignment of the ICM MG SB RAS (project 0315-2019-0003).

## 1 Hypersphere formula

The formula for a hypersphere can look like this, where $R$ is its radius:

$$R^2 = X^2+Y^2+Z^2+H^2 \tag{1}$$

then, after a small transformation, which consists in transferring the parameter $H$ along the fourth dimension to the left side of the equation, we can see on the right side of it a parametrically defined ball.

### 1.1 Ball formula

The formula for a parametrically defined ball can look like this:

$$R^2-H^2 = X^2+Y^2+Z^2 \tag{2}$$

* Corresponding author: kim@ooi.sscc.ru

https://doi.org/10.1051/e3sconf/202022302005

Thus, each point on the hypersphere corresponds to an interior point on one of two balls glued along the outer surfaces, depending on whether the parameter $-R \leq H \leq +R$ is positive or negative.

### *1.1.1 Interaction of homeomorphic varieties*

The diagram of such correspondence, which explains the interaction of homeomorphic manifolds, is shown in Fig. 1, for two-dimensional spaces of smaller dimension.

**Fig. 1.** The geographic map and the globe are informationally the same system, but visualized in different spaces

On the left side of the figure, the maps of the northern and southern hemispheres are glued together along the equator to form a flat, two-sided map. Returning to the hypersphere, we note that the centers of two generated balls are the poles of the hypersphere and, accordingly, the radii will form the guiding "meridians" of the hypersphere, like a dandelion inflorescence in Fig. 2. The section of the meridian with a parallels of subsphere selected inside an arbitrary ball uniquely determines the location of the point both inside this ball and on the surface of the hypersphere.

**Fig. 2.** The layout of the meridians in the form of the structure of a dandelion inflorescence

### *1.1.2 Cartographic grid*

So, the coordinate of a point will be determined by a set of geographic coordinates (*hyperparallel H, natural parallel, natural median*) - that is, a point on the hypersurface of the hypersphere will be determined by three coordinates, while a continuous line drawn in the ball is also continuously displayed on the surface of the hypersphere.

### *1.1.3 Four-dimensional shapes*

The issues of the geometric representation of multidimensionality began to be given scientific attention from the middle of the 19th century, which was soon reflected in Einstein's four-dimensional space-time concept, linking the relationship between space and time. So, according to the theory of relativity, with the simultaneous beginning of motion any two arbitrary objects from starting point to one target point with different speeds and trajectories in space, the durations of the times spent will differ when they meet. Modern theoretical physicists are currently developing an unusual physical theory of strings that works only in multidimensional space, dimensions from 10 or 26 dimensions.

The simplest geometric body for an n-dimensional space is a simplex, which is a convex polyhedron with the number of vertices which are per one unit more than the dimension of the space itself. Opposite any vertex of the simplex lies a face containing all the other vertices. So for a zero-dimensional space, a simplex is represented by a point. For a one-dimensional space, a simplex is already a line segment. In two-dimensional space, a simplex looks like a regular triangle, and in three-dimensional space, a simplex becomes a tetrahedron. Let's represent the described structures in a geometric manner. For example, a point represents a simplex of zero-dimensional space. By "pulling out" another point from a first point and connecting them with a line segment of length "one", we obtain a simplex of one-dimensional space. Building a regular triangle on a line segment, we obtain a simplex of two-dimensional space. Transforming a regular triangle into a regular tetrahedron, we represent a simplex of three-dimensional space. From geometric considerations, it is possible to calculate the radii of the spheres circumscribed around the "unit" simplexes:

$$r0 = 0$$

for a sphere in zero-dimensional space,

$$r1 = 1/2$$

for a sphere in one-dimensional space,

$$r2 = 2 * sqrt\ (1-r1^2)\ /\ 3$$

for a sphere in two-dimensional space and finally

$$r3 = 3 * sqrt\ (1-r2^2)\ /\ 4$$

for a sphere in three-dimensional space. The algorithm for constructing the subsequent simplex is as follows. In the presence of an n-dimensional simplex, with n vertices, the addition of the next vertex is postponed orthogonally to the face of the n-dimensional simplex at a distance calculated by the Pythagorean theorem with the hypotenuse "unit" side length and the known mathematical cathetus of the circumscribed sphere radius calculated in the previous step. The center of mass of the system is calculated on the constructed cathetus from the ratio *1:n*, which allows you to calculate the length of the next radius using the formula

$$r_n = n * sqrt\ (1-r_{n-1}^2)\ /\ (n + 1).$$

Calculating the passage to the infinity limit, we obtain the key constant:

$$r = 1\ /\ sqrt\ (2)$$

In the work [4], infinite-dimensional spaces are considered, thus in this connection, it is possible to derive an iterative formula for calculating geographic coordinates when passing to spaces of higher dimension - *(H..., (H3, (H2, (H1, parallel, meridian))))*. Let's see what the simplex framework looks like on a hypersphere. Any five points inside a 3-D ball sets wireframe points on the hypersphere, forming a simplex-five-cell.

### *1.1.4 Matryoshka structures*

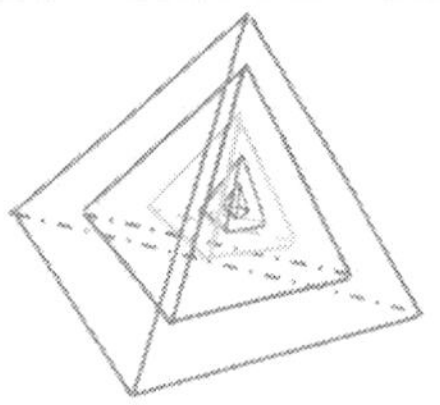

**Fig. 3.** Nested tetrahedron

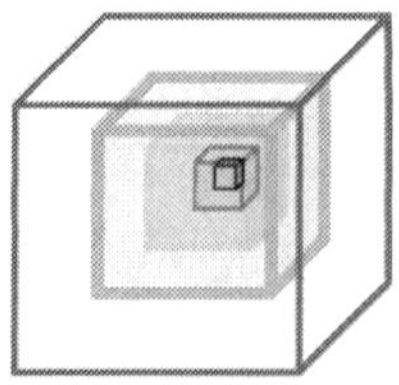

**Fig. 4.** Nested Cubes

"Matryoshka structures" [5], allow a cartographic way to "see" the hyper-relief in four-dimensional space. The approach is based on the expansion of the cartographic representation of the relief in the form of level contours in the direction of structures nested in three-dimensional space that correspond to one "height" along the fourth dimension (Figs. 3-4). Linear marking of the volume elements of a regular Platonic polyhedron with zero value on its boundary surface and values equal to the minimum distance to the boundary, which specify the height in the fourth dimension, creates nested structures in the form of regular Platonic solids, forming five four-dimensional pyramids, with their three-dimensional bases. The nested sphere markup generates a *4D* cone

$$R^2 = X^2 + Y^2 + Z^2 = (R-h)^2.$$

These primitives will form the basis for constructing a prototype design hyper-relief on a hyper-sphere, creating the prerequisites for creating a GIS for a four-dimensional space, which allows solving typical geographic information problems like of determining "visibility zones". The applicability of the "natural clustering of multidimensional space" developed in the laboratory was tested for economic problems with multifactor data, and was reported at a number of conferences.

## 2 Bulbous layers of the interior of the ball

A cartographic grid is an image on a map of geographic meridians and parallels in a particular cartographic projection. Serves for building a cartographic image and allows you to determine the coordinates of points on the map. When using a map of such a grid, it is possible to determine the coordinates of any point (geographic or rectangular, depending on the type of grid) and azimuths of the lines, as well as to judge the magnitude of distortions of the cartographic projection in different parts of the map.

### 2.1 Cartographic film

Setting the adjacent cartographic surfaces of the spheres inside the ball with maps of changes in the geography of the Earth in time, we get layer-by-layer parallels in four-dimensional space, which can be tracked in the form of a motion picture showing the process of changing the world (Fig. 3), sequentially moving to the next geographical map in the order of layers. At the same time, on the surface of the hypersphere, the graphic image will not be directly related to the existing film display formats, and it can be considered that we are talking about a different reading or another way of visualizing four-dimensional information.

**Fig. 5.** The dynamic development in time of the cartographic image of the Earth can be represented by multilayer schemes of its changes.

### 2.2 Obstacles

According to the expertise of the Federal Service for Intellectual Property of the FEDERAL INSTITUTE OF INDUSTRIAL PROPERTY of Russia (No. 2016138924/12 (062081) dated 01/30/2018), the concept of a four-dimensional body (object) cannot be recognized as relating to the object of the invention, since it characterizes only a scientific theory not confirmed by an official source of the RAS, in while an important global indicator of a newly forming technology trend, the US patent should be highlighted. Yao, 01/31/2017, United States Patent 9554776 "Method for adjusting ROI and 3D/4D imaging apparatus using the same", which presents a device that creates a virtual 4D object. Another example of the development of the latest information technologies is the 4D-Anatomy virtual textbook - a cloud-based, interactive, resource for modeling the dissection of a physical body, as well as a modern educational platform. The creators of the program have implemented a powerful interactive resource on human anatomy. You get not just an image, but you can rotate it and even see from different angles how the human heart beats [3]. It is easy to imagine how useful such opportunities are in cardiology, in the face of increasing global attention to the ever-growing role of the digital economy.

## 3 Conclusion

If we recall that photography, appeared more than a century and a half ago, which was based on a camera obscura, and was always perceived as an exclusively two-dimensional object, was developed into the modern film industry, which gave rise to a much more complex object than simple “two-dimensionality”. The desire to go beyond the plane of the screen, prompted the inventors to stereocinematography - a variety of cinematic systems that imitate the presence of a third dimension, or cause the viewer to have the illusion of the depth of space. It is based on the phenomenon of human binocular vision and the optical parallax effect. Modern computer technologies make it possible to create pseudo-stereo images using computer graphics, without using stereo cameras. Similarly, it is possible to convert an existing "flat" image into three-dimensional by synthesizing the second part of a stereo pair. The terms “three-dimensional graphics” (3D-graphics) and “3D-cinema” used on the Internet describe fundamentally different phenomena and technologies. With the help of modern computer technology, there is no need to be fixed on a flat screen, while the dynamics of the process can be produced in volume space. And here the "impossibility to look behind the screen" is no longer an obstacle. Thus, holographic objects under development will certainly lead us to the four-dimensional world of 4D cinematography, which, due to human inertia, is still referred to as a marketing term.

## References

1. Poincaré's conjecture. [Electronic resource] URL: https://https://en.wikipedia.org/wiki/Poincaré_conjecture (the date of the reference search: 24/10/2020).
2. P. A. Kim , P. A. Kalantaev Four-dimensinal visualization //The 11th International Conference "Pattern recognition and image analysis: new information technologies"

PRIA-11-2013, Samara, The Russian Federation, September 23-28, 2013,Vol.II, p.604-606

3. 4D Anatomy [Electronic resource] URL: https://www.4danatomy.com/ (the date of the reference search: 09.07.2020).
4. P.A. Kim About the radius of the circumscribed sphere of the infinite-dimensional simplex //The XIX International Conference on Science and Technology Russia-KoreaCIS. August 29–31, 2019. Moscow.Proceedings p.111-114.
5. Kim P.A. SCALABLE HYPERRELIEF MODEL IN 4D-SPACE // Proceedings of the VIII International Scientific Congress.Interexpo <GEO-Siberia-2012>, April 10-20, 2012, Novosibirsk, Russia, International Scientific Conference <Dist. met. probe.Earth and photogram., monitor. environment, geoecology>: collection of materials in 2 volumes, v. 1. -Novosibirsk: SGGA, 2012.-P.171-175.

# Building a model for predicting the water level in a river using remote sensing data from open sources

*Alexey Kolesnikov*[1,*], *Pavel Kikin*[2] , and *Anastasia Nungesser*[1]

[1] Siberian State University of Geosystems and Technologies, Department of Cartography and Geoinformatics, 630108 Plakhotnogo street 10, Novosibirsk, Russia
[2] Peter the Great St. Petersburg Polytechnic University, Higher School of Theoretical Mechanics, 195251 Polytechnicheskaya Street 29, Saint Petersburg, Russia

**Abstract.** The article discusses the possibility of predicting the water surface area of a river (and based on these values, the calculation of the water level) based on only open data of remote sensing. The area and depth of snow cover, the intensity of precipitation according to Landsat and Sentinel data and the monitoring indicators MODIS, Copernicus, REMSS are used as initial parameters. For the selected parameters, the degree of influence on the final forecast was assessed.

## 1 Introduction

Summer-autumn floods occurring on rivers can cause serious damage to industrial facilities, settlements, agricultural land. Therefore, their forecasting is a complex multi-level task, the relevance of which is due to the current state of economic systems. This primarily applies to urbanized and industrial regions. [1-3]. Since now a large amount of various remote sensing data is presented in the open access on a time interval of more than 10 years (with a fairly high frequency), it becomes possible to study almost any territory in terms of hydrographic objects and their relationships with climatic factors [4,5]. When building a forecast, it is very important and relevant to choose the appropriate research method. In hydrology, various methods of forecasting are widespread: graphic-analytical, water-balance, mathematical modeling, etc. There are a number of approaches for solving this problem, both deterministic and stochastic. In this study, machine learning methods were chosen to construct a mathematical model of the river flood, so it was necessary to carefully prepare the data set in order to "train" the relevant algorithms.

Today's remote sensing open data sources make it possible to collect dozens of environmental parameters every day, which, on the one hand, makes it possible to build the most comprehensive models, on the other hand, it can lead to unnecessary complication of the model and it's overfitting. [6,7]. Therefore, it was important to analyze the algorithms for assessing the importance (impact for the target feature prediction) of each of the

* Corresponding author: alexeykw@mail.ru

parameters to the final result and use them to analyze the initial data in the described problem.

## 2 Materials and Methods

In the very beginning of the study the sufficiency of the most logically grounded data was checked in order to construct a forecast of the water level and assess its accuracy. The territory of Irkutsk was chosen as the object of research. One of the most difficult, but often determining factors in the formation of runoff is the absorption of incoming water by the soil and precipitation. The latter are predicted only with a lead time of 2-3 days, or are replaced by their norm over the lead time period. The amount of runoff during the flood is determined by three main factors: the amount of water reserves in the snow cover accumulated by the time of the forecast, the amount of precipitation from the moment of making the forecast until the expiration of its lead time, the water-absorbing capacity of the basin, which depends on the depth of soil freezing, its humidity in autumn and during the winter thaws, negative air temperatures at the Earth's surface and the snow reserves themselves. These factors determine the moisture absorption capacity of different parts of the basin in different ways, depending on the composition of the underlying surface and the characteristics of a particular year. As the initial data, we used data on snow cover in areas of mountain ranges from which the flow of Lake Baikal and the Angara River is formed (MODIS), information on precipitation on the territory of catchment areas (Copernicus, REMSS), as well as water surface parameters based on the interpretation of satellite images (Landsat, Sentinel). The collected data from open sources were transformed in such a way that it was possible to form a general table and, on the basis of which, build a mathematical forecasting model and evaluate its quality (the execution scheme is shown in Figure 1).

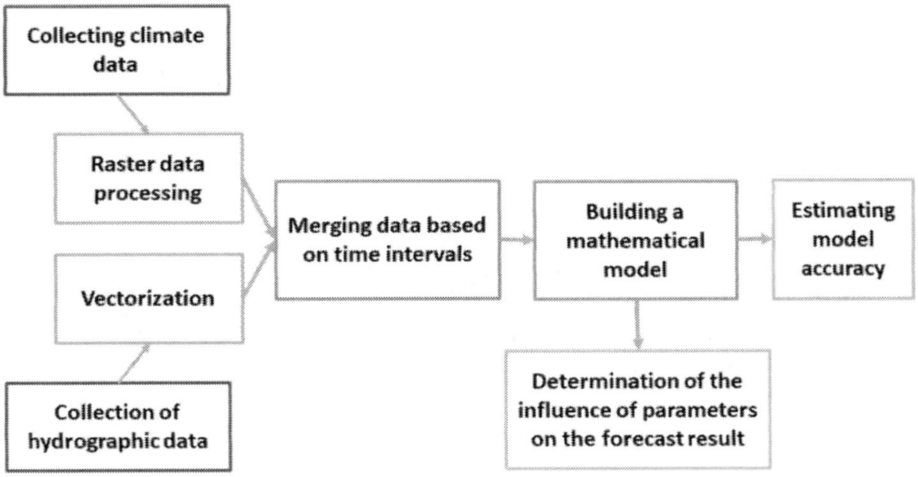


**Fig. 1.** Study execution pipeline

At the stages of setting a machine learning problem and generating data, it is not always clear which features are important for constructing an optimal algorithm, therefore, data often contains a lot of redundant (noise) information. The appearance of noise signs degrades the quality of the algorithm and slows down its work. Therefore, in most cases, before solving the problem of classification, regression or forecasting, it is necessary to select those features that are most informative. Choosing the right features can be a more significant task than reducing data processing time or improving classification accuracy.

Therefore, a block for determining the importance of individual features was added to the study. In addition to demonstrating the importance of features directly, one can evaluate the general concept of the model, its interpretability. Interpreting methods for machine learning models are divided into local and global. Global methods are designed to show which factors in general have the greatest influence on the structure of the model and on its predictions. Local methods try to explain how a given prediction was made. Often local methods can be used as a basis for a more global interpretation, for example, through averaging or visualization. Local methods LIME and SHAP were used for the assessment.
Data were collected for certain areas (Figure 2), the location of which was selected based on the hydrological description of the Angara River and Lake Baikal, for which the main source of water is the Selenga River. The time interval from 2014-2019 was taken for the analysis.
For remote study of the areas of the water surface of lakes according to space data, the following methods of processing satellite images are widely used - determination of multichannel spectral indices (water indices), thematic classification with training, linear separation, single-channel classification using the separation threshold. The paper considers a method for decoding water surfaces using multi-channel spectral indices MNDWI - Modified Normalized Difference Water Index. MNDWI is determined using 3 and 6 channels of Landsat 8 OLI and is currently the most common index for identifying surface water objects in satellite imagery. The boundary of the definition of water surfaces passes through zero, the values of the indices greater than zero correspond to water surfaces.

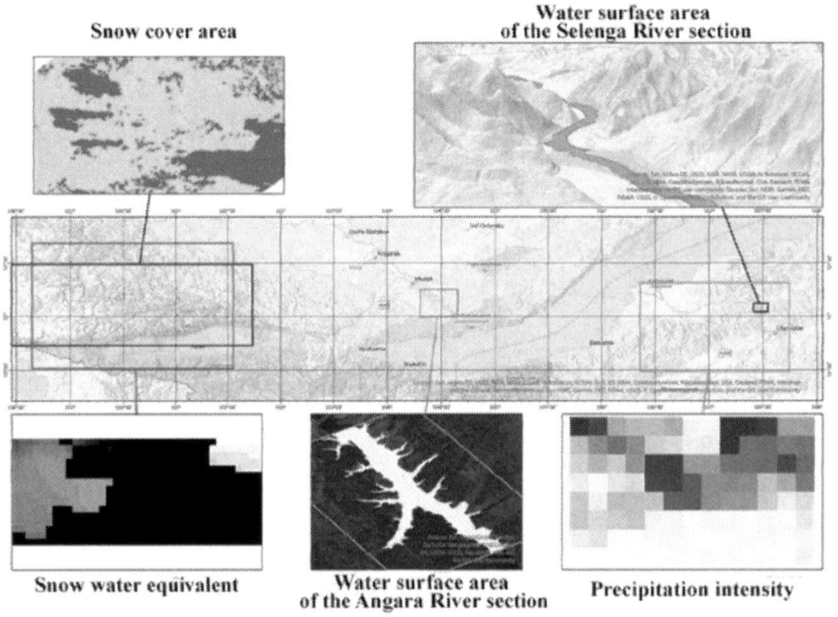


**Fig. 2.** Data and collection sites

Since most of the training sample is presented in raster form, it was necessary to choose a way to represent each raster in the form of one row of the table (for comparison with the corresponding value of the water cover area). The simplest option is to representing each band of a raster image as a flat array. But it was not suitable due to the significant increase in the dataset. In this regard, it was decided to calculate summary quantitative indicators for each raster: maximum, minimum, average, median, standard deviation. An important step when using multispectral images is to eliminate the redundancy of such images while preserving important spectral information. In this regard, considerable attention of researchers was paid to both methods of controlled selection of features and uncontrolled methods of reducing the dimension of data. While both of these directions inevitably lead to some loss of spectral information, successful results in terms of the quality of subsequent classification have been demonstrated in both directions. For a number of objective reasons, such as high computational costs, less resistance to scene changes, the need for a classified set, feature selection methods are inferior to dimensionality reduction methods. And, although when using data dimensionality reduction methods, the original interpretation of the data (spectral signatures) is lost, the latter are more often used when working with multispectral images [10-12]. In order to at least partially preserve the information about the spatial arrangement of pixels of the original rasters, the PCA algorithm was used for each of the rasters. The first 5 components for each of the image channels were extracted into a pivot table.

The ArcGIS Model Builder tool was used as an automation tool. For each raster, cropping was performed according to the data collection area and statistical indicators were calculated, which were saved to a CSV file. This approach made it possible to reduce the time for processing each raster separately, but a fairly large part of the operations (forming datasets for loading, placing them on local disks, combining the results into one table) still had to be done manually. As a result, for further experiments, it is planned to either extend the ArcGIS scripts by writing additional functions in Python, or to transfer most of the GIS operations to some kind of spatial database, since their functionality now allows this. It should also be noted that Model Builder does not always use RAM efficiently, and on datasets containing hundreds of elements, the processing speed at each iteration gradually decreases.

## 3 Results

The importance of individual parameters was assessed based on the calculation of the correlation and the rating of the contribution of each parameter to the final forecast using the LIME [13] and SHAP [14] algorithm, as well as the Feature Importance mechanism built into the random forest and decision tree model of scikit learn" library [15]. According to the assessment results, the area of snow cover received the highest impact rating.

To construct mathematical models, linear regression, decision tree and random forest algorithms were selected. The choice is due to the fact that these algorithms are computationally simple and their results are well interpreted. The root mean square error error was chosen as the criterion for assessing the accuracy.

As a result, the random forest algorithm showed the highest accuracy, which for the Angara River section was 0.31 sq. km. And for the section of the Selenga River 0.3 sq. km., which is about 10% of the deviation from the true value.

## 4 Discussion

To improve the accuracy of the forecast, it is planned to increase the number and areas of data collection regions, which will allow analyzing the effectiveness of the choice of locations and time intervals for collecting data on snow cover and formalizing the selection principles based on an open digital elevation model and basic hydrological analysis. Also, the quality can be improved by creating ensembles of standard machine learning algorithms together with time series forecasting algorithms (such as SARIMAX or LSTM).

Also, since the resulting value is only a digit, you need to choose an algorithm for forming the polygon based on the relief data and the calculated area value.

## References

1. M. Norwaziah, P. Nur, I. Izleen, M. Umi, J. Siti, O. Nurul, Proceedings of the Second International Conference on the Future of ASEAN (ICoFA) *Forecasting River Water Level Using Artificial Neural Network*, **2** (2018), 10.1007/978-981-10-8471-3_41.
2. V. Hung, L. Xuan Hien, H. Tuan, XVII World Water Congress of International Water Resources Association. *Using Machine Learning to forecast river water levels in tidal areas for operating sluice gates in Hai Duong province, Vietnam*. (2020)
3. H. Dinna, N. Hardini, L. Dinda, International Conferences on Information System and Technology. *Forecasting River Water Quality using Autoregressive Integrated Moving Average (ARIMA)*, 158-163. (2019) 10.5220/0009907201580163.
4. L. Plastinin, V. Stupin, B. Olzoev, Digital geography: proceedings of the All-Russian scientific-practical conference with international participation. *Integration of the regional cartographic and space monitoring system into the geo-information environment,* **1,** (2020)
5. D.V. Andreev, The successes of modern natural science. *Application of gis technologies to determine flooding in the republic of Sakha (Yakutia).* **11**, 43-47, (2019)
6. W.E. Chapman, A.C. Subramanian, L. Monache, S.P. Xie, F.M. Ralph, Geophysical Research Letters. *Improving Atmospheric River Forecasts With Machine Learning.* **46(1),** (2019) 10.1029/2019GL083662.
7. C. Tarau, K. Lee, W. Anderson, C. Morrison, T. Hendrick, International Conference on Environmental Systems (ICES 2020), *Thermal Concept for Planetary Ice Melting Probe*, (2020)
8. M.A. Golyatina, K.A. Kurganovich, Materials of Intern. Conf. Geosystems and their components in Northeast Asia: evolution and dynamics of natural, natural-resource and socio-economic relations, *The study of the change in the number of lakes in the Transbaikal Territory with the use of MNDWI*, 174-179, (2016)
9. G.L. Feyisa, H. Meilby, R. Fensholt, S.R. Proud, Remote Sensing of Environment, *Automated Water Extraction Index: A new technique for surface water mapping using Landsat imagery*, **140**, 23-35, (2014)
10. J. Wang, C.-I. Chang, IEEE Trans. Geosci. Remote Sens. *Independent component analysis-based dimensionality reduction with applications in hyperspectral image analysis,* **44 (6)**, 1586-1600, (2006)
11. Z. Zhang, Z. Hongyuan, SIAM Journal on Scientific Computing, *Principal Manifolds and Nonlinear Dimension Reduction via Local Tangent Space Alignment*, **26(1)**, 313-338, (2005)
12. C.M. Bachmann, T.L. Ainsworth, R.A. Fusina, IEEE Trans. Geosci. Remote Sens., *Improved Manifold Coordinate Representations of Large-Scale Hyperspectral Scenes*, **44(10)**, 2786-2803, (2006)
13. R. Marco Tulio, S. Singh, C. Guestrin. Proceedings of the 22nd ACM SIGKDD international conference on knowledge discovery and data mining. ACM, *Why should I trust you?: Explaining the predictions of any classifier*, (2016)
14. S. Lundberg, S. Lee, Advances in Neural Information Processing Systems (NIPS 2017), *A Unified Approach to Interpreting Model Predictions*, **30,** 4765—4774, (2017)
15. J. Demsar, T. Curk, A. Erjavec, C. Gorup, T. Hocevar, M. Milutinovic, M. Mozina, M. Polajnar, M. Toplak, A. Starc, M. Stajdohar, L. Umek, L. Zagar, J. Zbontar, M. Zitnik,

B. Zupan, Journal of Machine Learning Research. *Orange: Data Mining Toolbox in Python* 2349–2353, (2013)

# Methodology for Developing Algorithms for Compressing Hyperspectral Aerospace Images used on Board Spacecraft

*Assiya* Sarinova[1], *Alexander* Zamyatin[2]

[1]Toraighyrov university, Electrical Engineering and Automation Department, 140000 Lomov street 64, Kazakhstan
[2]National research Tomsk state university, Institute of Applied Mathematics and Computer Science, 634050 Lenina avenu 36, Russia

**Abstract.** The paper describes a method for constructing and developing algorithms for compressing hyperspectral aerospace images (AI) of hardware implementation for subsequent use in remote sensing Systems (RSS). The developed compression methods based on differential and discrete transformations are proposed as compression algorithms necessary for reducing the amount of transmitted information. The paper considers a method for developing compression algorithms, which is used to develop an adaptive algorithm for compressing hyperspectral AI using programmable devices. Studies have shown that the proposed algorithms have sufficient efficiency for use and can be applied on Board spacecraft when transmitting hyperspectral remote sensing data in conditions of limited buffer memory capacity and communication channel bandwidth.

## Introduction

Hyperspectral remote sensing AI is important for observing and studying changes in the Earth's surface, monitoring natural resources and the consequences of emergencies, etc. Currently, the development of software systems for transmitting such data is an urgent task. In solving this problem, there are two areas of research: the development of compression algorithms used in ground-based remote sensing data reception and processing centers; and those used on Board SPACECRAFT. Research is actively conducted in the development of compression algorithms of the first direction, in which there are many publications [1-10, 12]. In the second direction of research, there is a potential for developing algorithms, dictated by the necessary list of problems for solving the compression problem.

It is proposed to formulate some problems of developing algorithms for compressing hyperspectral AI that are applicable on Board.

1. The amount of incoming data. Hyperspectral AI has a range of several hundred and thousands of spectral channels. At the same time, the size of channels reaches hundreds of thousands of pixels.

2. Memory resources. The technical characteristics of the hardware that receives, stores, and transmits such data are limited in memory and processing power.

3. Data transmission. Huge volumes are formed on Board the SPACECRAFT at a certain speed exceeding the capabilities of the transmitted communication channel, which has a fixed bandwidth that is insufficient for transmitting uncompressed hyperspectral AI.

4. Data quality. Data quality requirements are very high, as the incoming information is unique.

Compression cannot be applied under these conditions. The problems listed above that arise when developing software systems that are applicable on Board the SPACECRAFT give rise to the following range of requirements for the compression algorithm:

- the speed of a stream of compressed data;
- low computational complexity;
- high compression ratio;
- error control for lossy compression.

Currently, well-known algorithms based on truncated block encoding [1], differential pulse modulation [1], discrete cosine transform [2], and discrete wavelet transform [2] exist and are used for compressing hyperspectral AI. These algorithms on the SC bot, which require large computing resources, do not always meet the above compression requirements. Therefore, it is necessary to develop new methods for compressing hyperspectral AI, which are not inferior in their efficiency to the known methods, but require less computing resources.

Thus, in this paper, we propose to consider a method for developing algorithms for compressing hyperspectral AI that meet these requirements.

## 1 Description of the methodology for developing compression algorithms

The methodology for constructing the development of an algorithm for the hardware implementation of compression should be divided into three stages.

The first stage is the selection and justification of hardware and software for the compression of hyperspectral AI.

The second stage is the development of lossless and lossless compression algorithms on programmable integrated circuits.

The third stage is testing and conducting experiments of the developed algorithms [11,13] in comparison with the known algorithms in terms of performance, speed and efficiency in compression ratio.

**Stage 1.** To test the developed algorithms at the first stage, a 32-bit STM32 microcircuit from Electronics and CocoxIDE software with embedded C ++ language was chosen as a tested programmable integrated circuit.

Some of the latest processors for embedded systems are those based on the ARM Cortex-M3 architecture. These processors are designed for use in Digital Signal Processing (DSP).

General description of ARM architecture and 32-bit STM microcontrollers. ARM processors are a key component in many successful 32-bit embedded systems. ARM processors are widely used in mobile phones, tablets and other portable devices. ARMs are based on RISC architecture, which reduces processor power consumption and thus makes them ideal for embedded systems.

*Benefits of using:*

1. Variable number of execution cycles for simple instructions. Simple ARM instructions require more than one cycle to execute. For example, the execution of the Load and Save instructions depends on the number of registers that are passed to them.

2. Ability to combine commands of shift and rotation with commands for information processing.

3. Conditional execution - the statement is executed only if a specific condition is met. This improves performance and eliminates branching statements.

4. Enhanced instructions - ARM processors support enhanced DSP instructions for digital signal operations.

There are two software products that provide functionality for executing firmware without using a development environment: STM32 ST-LINK Utility and ST Visual Programmer.

STM32 ST-LINK Utility is designed to work with 32-bit controllers via ST-LINK interface.

The general characteristics of the core of STM32 microcontrollers are presented in table 1.

**Table 1.** Main characteristics of the core of STM32 microcontrollers

| Characteristic | *Value* |
|---|---|
| data word width, bit | 32 |
| architecture | harvard |
| conveyor | 3- stepped |
| set of instructions | RISC |
| organization of program memory, category | 32 |
| prefetch buffer, bit | 2x64 |
| average instruction size, bytes | 2 |
| interrupt type | vectorized |
| interrupt response delay | 12 cycles |
| power management modes | sleep, sleep on exit, deep sleep |
| debug interface | ST-LINK, JTAG |

Microcontrollers of this type are built on the Harvard architecture and have a 3-stage pipeline that minimizes command execution time. They are designed to build systems with maximum energy efficiency and have multiple power management modes. They use internal memory interfaces that are wider than the average instruction length. This minimizes the number of accesses to the memory bus, and hence the power consumption associated with bus operations and non-volatile memory reads. Continuous interrupt processing technology with the exception of internal stack operations reduces the response time to interrupts and eliminates unnecessary stack operations.

The STM32F4 Discovery Board is equipped with:

- A stm32f407vgt6 microcontroller with a Cortex-M3 core clocked at 168 MHz, 1 MB of Flash memory, 192 KB of RAM;
- ST-Link debugger for debugging and programming;
- power the Board via USB or an external 5V power supply;
- ST MEMS LIS302DL motion sensor and digital accelerometer outputs;
- sound sensor MP45DT02, ST MEMS;
- CS43L22 audio DAC;
- eight LEDs;
- two buttons (for user programming and for restarting).

Thus, the debugging Board is equipped with a large number of peripherals, which allows you to implement algorithms of varying complexity on it.

2. Stage. An original and efficient algorithm for lossless compression of hyperspectral AI by regression transformation is developed. An algorithm for lossless compression of hyperspectral aerospace images, characterized by the use of channel-difference linear regression transformation, which significantly reduces the range of data changes and increases the compression rate due to this. The main idea of the transformation is an

algorithm that finds pairs of correlated channels and then creates lossless transformed blocks using regressin analysis, which reduces the size of the aerospace image channels and converts them before compressing the modified Huffman algorithm.

Several orthogonal transformations adapted for lossy compression of hyperpectral AI have been developed. An adaptive algorithm for discrete-cosine transformation with subsequent quantization with a loss level and compression by Huffman encoding is developed.

Adapted Huffman algorithm. The well-known standard Huffman code table is recommended for JPEG images, not for hyperspectral AI, so the generated one is suggested for encoding. Modification of the table is as follows:

1) each possible pair of RZ is assigned a natural number intMerge starting with two;

2) consider the binary notation of the number intMerge and pay attention to its length (lenMerge). For example, for (R;Z)=(4;7) intMerge = 117, binary entry -1110101, lenMerge = 7;

3) let's build the code as follows: take one (lenMerge - 1) times, assign 0 and the binary entry of the number intMerge without the leading unit. For clarity, after zero, we will add an apostrophe: 1111110'110101.

For an example from the table: (R;Z) = (14; 1) intMerge = 17, binary entry-10001, lenMerge = 5. Huffman Code: 11110 ' 0001.

The number Z does not exceed 14, since we replace the sequence of 15 zeros with the code 111111110 ' 00000000 (as if intMerge was 256). The code for the end of the block is 00. The pair (R;Z)=(15;0) is encoded as 01. If R exceeds 16, we assign intMerge, starting from 257, in the same way.

This modification made it possible to increase the compression rate of hyperspectral AI by analyzing the frequencies of the R and Z pairs, since codes of shorter length are allocated to more common values during encoding.

At the second stage, an algorithm for multithreaded compression processing was developed and organized to improve the computational efficiency of compression.

Algorithm for multiple streams of hyperspectral AI encoding.

1. determining the number of encoded pairs in the set of ordered pairs (CBM);
2. the first thread receives and encodes the main generating file;
3. the second and subsequent streams encode subsequent pairs in the CBM;
4. upon release, all subsequent streams receive the next pair of channels until the CBM is empty;
5. when encoding is complete, we release all threads.

The algorithm is a single thread during decoding.

1. consider the next pair in the CBM. Let it be k1 and k2, where k1 is the generating file and k2 is the regressed file;
2. if the file k1 is the main generating file, or it is already decoded, then we decode the pair (k1- k2);
3. if the file k1 has not been decoded yet (it may be in the process), then we consider the next pair after k1, k2 in the CBM (without discarding it);
4. go to step 2 – if such a pair is found, go to standby mode-if not.
5. at the end of decoding, we notify the rest of the threads in waiting, go to step 1 (if the CBM is not empty), or complete the execution of the thread). In this case, we discard the encoded pair (k1 - k2) from the CBM.

## 2 Experiments of the developed hardware implementation algorithms

 https://doi.org/10.1051/e3sconf/202022302007

To determine the effectiveness of the proposed adaptive algorithm in terms of compression ratio and computational efficiency, as well as the limits of its applicability, a number of experiments were performed on hyperspectral AI (Aviris remote sensing system), (Fig.1).

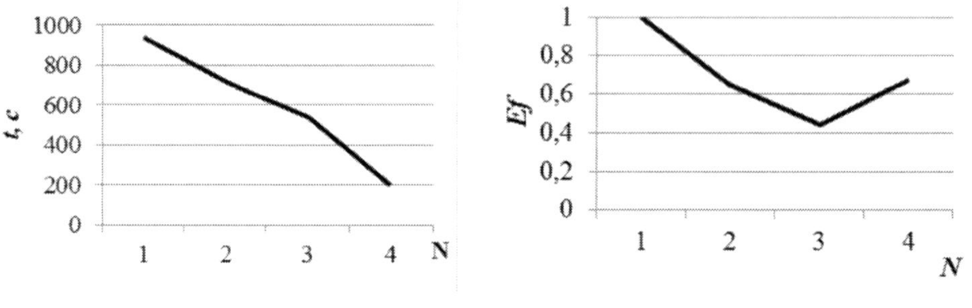


**Fig. 1.** Adapted compression algorithm for hyperspectral AI:
a) calculation time; b) execution efficiency

The experiments were performed on a PC with an IntelCore i7, 2.5 GHz processor and 4 GB of RAM running the Windows 10 operating system

JPEG Lossless/ Lossy compression algorithms, 7z archiver, and Winrar were used as analogs. The results of compression experiments using the proposed method are presented in tables 2-3. For comparison, the compression ratio and processing speed averaged over all experiments are taken

**Table 2.** Comparison of lossless compression algorithms

| *Compression algorithm* | *Compression ratio* | *Calculation time, s* |
|---|---|---|
| JPEG Lossless | 3,5 | 221452 |
| 7z | 3,8 | 205146 |
| Winrar | 3,9 | 220738 |
| Adaptive algorithm | 5,9 | 74816 |

**Table 3.** Comparison of lossy compression algorithms

| *Compression algorithm* | *Compression ratio* | *Calculation time, s* |
|---|---|---|
| JPEG Lossy | 5,7 | 225672 |
| Adaptive algorithm | 9,2 | 78235 |

It should be noted that in these experiments, the adaptive algorithm shows results in a compression ratio higher than the known solutions by more than 55 %, and significantly exceeds the processing speed by 3 times.

## Conclusion

Based on the results of the study, it is shown that the proposed method for developing lossless and lossy compression algorithms based on hardware implementation is highly effective in the compression ratio and processing speed. This is due to the fact that the algorithms were adapted for this implementation in order to be able to be used on Board the spacecraft. This method of constructing algorithms has a high speed due to the transition to hardware that operates with bits, providing an increase in speed at times.

Adaptive algorithms can be used both in remote sensing systems and in any other applications for processing and compressing hyperspectral AI.

References:
1. Gonsales R., Vuds R. Cifrovaya obrabotka izobrazhenij. - M.: Tekhnosfera, 2012. – pp. 55-67. (In Russian).
2. Kashkin V. B., Suhinin A. I. Cifrovaya obrabotka aerokosmicheskih izobrazhenij // Krasnoyarsk: SFU. 2008. 278 p. (In Russian).
3. Gashnikov, M.V. Bortovaya obrabotka giperspektral'nyh dannyh v sistemah distancionnogo zondirovaniya Zemli na osnove ierarhicheskoj kompressii / M.V. Gashnikov, N.I. Glumov // *Komp'yuternaya optika*. – 2016. – vol. 40, no 4. – pp. 543–551. (In Russian).
4. Petrov E.P., Harina N.L., Suhih P.N. Metod szhatiya mnogorazryadnyh sputnikovyh snimkov bez poter'. Sovremennye problemy distancionnogo zondirovaniya Zemli iz kosmosa. 2016. vol. 13. no 2. pp. 203–210. (In Russian).
5. Fernando García-Vílchez, Jordi Muñoz-Marí, Maciel Zortea, Ian Blanes, Vicente González-Ruiz, Gustavo Camps-Valls, Antonio Plaza. On the Impact of Lossy Compression on Hyperspectral Image Classification and Unmixing. *IEEE Geoscience and remote sensing letters*, vol. 8, no. 2, 2011.
6. Meena B. Vallakati and Dr. R. R. Sedamkar. Low Complexity DCT-based DSC approach for Hyperspectral Image Compression with Arithmetic Code. *IJCSI International Journal of Computer Science Issues,* vol. 9, Issue 5, no 1, September 2012.
7. Yongjian Nian, Mi He, and Jianwei Wan. Low-Complexity Compression Algorithm for Hyperspectral Images Based on Distributed Source Coding. Hindawi Publishing Corporation. *Mathematical Problems in Engineering*. Vol. 2013, Article ID 825673, 7 p.
8. Ganeshraj P. and Sivasankar A. Scalable Compression Method for Hyperspectral Images // *Research Journal of Engineering Science.* vol. 2(3), pp.1–5. 2013.
9. Diego Valsesia, Enrico Magli . A Novel Rate Control Algorithm for Onboard Predictive Coding of Multispectral and Hyperspectral Images. 2014.
10. Dr. S.M.Ramesh, P.Bharat, J.Anand, J.Anbu Selvan. Analysis of Lossy Hyperspectral Image Compression Techniques // *International Journal of Computer Science and Mobile Computing*, Vol.3 Issue.2, February – 2014, pp. 302–307.
11. A. Sarinova, A. Zamyatin. E3S Web Conf **149**, 02003 (2020).
12. S. Kudubayeva, N. Amangeldy, A. Sundetbayeva and A. Sarinova. *ACM International Conference Proceeding Series* **8**, (2019).
13. A. Sarinova, A. Zamyatin and P. Cabral. DYNA (Colombia) **82** (190), 166-172 (2015).

# Experimental evaluation of nonparametric clustering algorithms for image segmentation

*Yuriy* Sinyavskiy[*], *Sergey* Rylov , and *Igor* Pestunov

Federal Research Center for Information and Computational Technologies, 630090, Novosibirsk, Russia

**Abstract.** Experimental evaluation of 12 nonparametric clustering algorithms for image segmentation was made. Algorithms developed in FRC ICT are compared to ones from ENVI, ELKI and Smile software packages. Seven model datasets were generated to estimate clustering accuracy. The computational efficiency was evaluated using digital photographs and fragments of multispectral images obtained from WorldView-2 satellite.

## 1 Introduction

Image segmentation is required for solution of a number of applied problems. These can be multispectral images obtained from satellites, aircrafts or unmanned aerial vehicles, as well as conventional digital photographs (e.g. medical examinations data). Segmentation has two main goals – splitting the image into parts for further analysis, and grouping pixels into higher-level informative structures [1]. One of the most common approaches to image segmentation is based on the use of data clustering algorithms [2]. Image segmentation is usually performed with neither *a priori* information about the probabilistic characteristics of classes, nor training samples. In these conditions, the most suitable is nonparametric approach to clustering [3]. It allows detecting clusters of complex structure without strict restrictions on probability density function. However, it has not become widespread due to high computational complexity. The use of the grid-based approach makes it possible to achieve high computational efficiency due to processing relatively small number of cells instead of data elements. But the clustering accuracy of detected clusters strongly depends on the grid structure [4]. Efficient density- and grid-based algorithms for multispectral images segmentation have been developed in the FRC ICT. In this paper, an experimental comparison of these algorithms and six most popular clustering algorithms implemented in ENVI [5], ELKI [6] and Smile [7] software packages is performed.

## 2 Algorithms and datasets

Experimental evaluation was performed using twelve clustering algorithms. Six of ones (HCA_MS [8; 9], HECA_MS [10], ECCA [11], CCAE [11], MeanSC [12], EMeanSC [13])

* Corresponding author: yorikmail@gmail.com

has been developed at the FRC ICT. Three efficient nonparametric algorithms (DBSCAN [14], OPTICS [15] and DENCLUE [16]), were taken from ELKI and Smile software packages for data mining. In addition, a parallel implementation of the effective (in terms of clustering quality) iterative density-based MeanShift algorithm [17] was made. The number of iterations in the experiments was limited to ten. Furthermore, software implementations of $k$-means [18] and ISODATA [18] clustering algorithms from ENVI software package were evaluated (the number of iterations was also limited to ten; cluster centers from the previous iteration were used to initialize the next one, improving the quality of segmentation). These algorithms are included in most popular software packages for satellite image processing, and therefore are often used in practice.

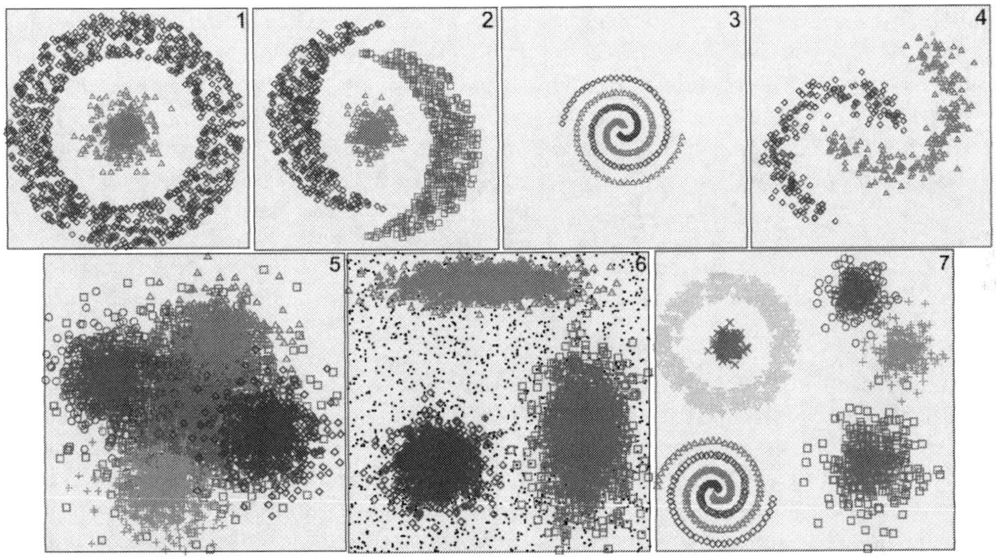

**Fig. 1.** Reference clustering for model datasets (black dots mark «noise»).

**Fig. 2.** Test images No. 1-5 (digital photos). Image sizes are 0.3, 1.0, 2.2, 5.0 and 13.8 millions of pixels, respectively.

**Fig. 3.** Test images No. 6-8 (fragments of multispectral images obtained from WorldView-2 satellite). Image sizes are 4.2, 9 and 12 millions of pixels, respectively.

In this work, seven model datasets (Figure 1) and eight test images were used – five digital photos (Figure 2) and three fragments of multispectral images obtained from the WorldView-2 satellite (Figure 3). Three spectral channels (red, green and blue) were used for processing digital photos, and four (red, green, blue and near-infrared) – for satellite images. The model datasets and RGB composites for the test images are available at [19]. The experiments were performed on a personal computer with an Intel Core i7 CPU (4 cores, 2.3 GHz each) and 8 GB RAM.

## 3 Experimental evaluation

In the first experiment, the clustering accuracy was estimated according to the following definition. Suppose that for a dataset $X = \{x_1, \dots, x_N\}$ of size $N$, the reference partition $g^*$ into $M$ classes $\{G_0^*, \dots, G_M^*\}$ is known. Then, for an arbitrary partition (clustering) $g$ into $K$ clusters $\{G_0, \dots, G_K\}$, the correspondence function $\gamma(G)\colon \{G_0^*, \dots, G_M^*\} \rightarrow \{G_0, \dots, G_K, \emptyset\}$, can be established performing condition $\forall(G \neq \bar{G})\colon \gamma(G) = \gamma(\bar{G}) \Leftrightarrow \gamma(G) = \emptyset$ and the highest value of the expression

$$n_\gamma = \sum_{i=1}^{M} [|G_i^* \cap \gamma(G_i^*)| \cdot I(\gamma(G_i^*) \neq \emptyset)],$$

where $I(\cdot)$ is the characteristic function. Then the clustering accuracy ($acc$) is determined as:

$$acc = \frac{n_\gamma}{N} * 100\%.$$

The goal of tuning the algorithm parameters was to obtain the maximum clustering accuracy. The MeanShift, $k$-means and DENCLUE algorithms do not allow detecting "noise", therefore, the $acc$ values for model dataset No. 6 were calculated without taking the "noise" class into account. The segmentation accuracy values obtained on the model datasets are presented in Table 1, and the processing time is shown in Table 2.

All of the algorithms, except for DENCLUE, $k$-means, and MeanShift, allow obtaining a reference partition for model datasets No. 1-3. Processing of model dataset No. 4 by the MeanSC and EMeanSC algorithms leads to one misclassified data element. The DBSCAN and OPTICS algorithms make it possible to obtain a slightly less accurate result. All algorithms, except for DBSCAN and OPTICS, allowed to obtain a clustering accuracy of about 85% for model dataset No. 5. Errors are caused by significant overlap of model classes. Model dataset No. 6 contains «noise» that could not be detected by the MeanShift, $k$-means and DENCLUE algorithms. The algorithms developed at FRC ICT made it possible to obtain a clustering accuracy higher than 84% for model dataset No. 6. The rest of the algorithms achieved less than 80% accuracy. When processing model dataset No. 7, only the HCA_MS, HECA_MS, ECCA, CCAE, MeanSC and EMeanSC algorithms allow to detect all classes. Applying DBSCAN, OPTICS, DENCLUE, and $k$-means algorithms leads to correct separation only for the normally distributed classes. The results of the experiment demonstrate that the algorithms developed at the FRC ICT are superior to the known nonparametric algorithms in terms of clustering quality and/or processing time.

In the second experiment, the considered algorithms were applied to test images and the processing time was compared. The results are presented in Table 3. Dashes in the table correspond to unacceptably high processing times (more than 18 hours). Analysis of the results shows that the DBSCAN, OPTICS and DENCLUE algorithms do not allow efficiently handle large images. In addition, their processing time increases significantly with increasing number of channels. Algorithms from the ENVI package are better adapted to image analysis, but the processing time for images larger than 9 million pixels exceeds 5 minutes. On the other hand, algorithms HCA_MS, HECA_MS, ECCA, CCAE, MeanSC and EMeanSC allow interactive segmentation of large multispectral images.

**Table 1.** Clustering accuracy (*acc*) for model datasets (percentage).

| Algorithm | Model dataset | | | | | | |
|---|---|---|---|---|---|---|---|
| | No. 1 | No. 2 | No. 3 | No. 4 | No. 5 | No. 6 | No. 7 |
| HCA_MS | 100 | 100 | 100 | 100 | 83.13 | 84.43 | 93.26 |
| ECCA | 100 | 100 | 100 | 100 | 84.83 | 85.65 | 98 |
| CCAE | 100 | 100 | 100 | 100 | 84.83 | 85.65 | 98 |
| HECA_MS | 100 | 100 | 100 | 100 | 85 | 86.57 | 93.26 |
| MeanSC | 100 | 100 | 100 | 99.75 | 86.7 | 89.18 | 98.6 |
| EMeanSC | 100 | 100 | 100 | 99.75 | 86.7 | 89.16 | 98.7 |
| MeanShift | 51.79 | 90.36 | 50 | 53.75 | 86 | 79.53 | 79.1 |
| $k$-means | 47.79 | 64.73 | 53 | 47.75 | 84.67 | 79.92 | 78.99 |
| DBSCAN | 100 | 100 | 100 | 99 | 65.2 | 84.56 | 90.63 |
| OPTICS | 100 | 100 | 100 | 99 | 65.2 | 84.56 | 90.63 |
| DENCLUE | 65.14 | 99.36 | 50 | 50.5 | 85.9 | 79.92 | 83.8 |

**Table 2.** Processing time for model datasets (in seconds).

| Algorithm | Model dataset | | | | | | |
|---|---|---|---|---|---|---|---|
| | No. 1 | № 2 | No. 1 | № 4 | No. 1 | № 6 | No. 1 |
| HCA_MS | 0.008 | 0.001 | 0.002 | 0.001 | 0.003 | 0.004 | 0.001 |
| ECCA | 0.003 | 0.005 | 0.003 | 0.003 | 0.001 | 0.028 | 0.013 |
| CCAE | 0.006 | 0.011 | 0.003 | 0.008 | 0.002 | 0.15 | 0.019 |
| HECA_MS | 0.006 | 0.008 | 0.004 | 0.003 | 0.006 | 0.03 | 0.021 |
| MeanSC | 0.002 | 0.005 | 0.004 | 0.002 | 0.01 | 0.014 | 0.017 |
| EMeanSC | 0.018 | 0.017 | 0.007 | 0.007 | 0.13 | 0.19 | 0.17 |
| MeanShift | 0.034 | 0.019 | 0.003 | 0.005 | 0.321 | 0.259 | 0.017 |
| $k$-means | 0.016 | 0.015 | 0.015 | 0.016 | 0.016 | 0.085 | 0.017 |
| DBSCAN | 0.031 | 0.016 | 0.016 | 0.016 | 0.374 | 1.17 | 0.031 |
| OPTICS | 0.09 | 0.074 | 0.048 | 0.065 | 1.131 | 2.461 | 0.14 |
| DENCLUE | 0.178 | 0.09 | 0.06 | 0.14 | 0.184 | 3.26 | 0.721 |

**Table 3.** Processing time for test images (in seconds).

| Test image | No. 1 | No. 2 | No. 3 | No. 4 | No. 5 | No. 6 | No. 7 | No. 8 |
|---|---|---|---|---|---|---|---|---|
| Image size (millions of pixels) | 0.3 | 1 | 2.2 | 5 | 13.8 | 4.2 | 9 | 12 |
| Number of channels | 3 | 3 | 3 | 3 | 3 | 4 | 4 | 4 |
| HCA_MS | 0.13 | 0.33 | 1.04 | 0.85 | 4.75 | 26.3 | 6.1 | 666 |
| ECCA | 0.17 | 0.48 | 0.66 | 3.1 | 5.5 | 1.3 | 3.54 | 3 |
| CCAE | 0.05 | 0.09 | 0.15 | 0.32 | 0.63 | 7.8 | 6.5 | 1 |
| HECA_MS | 0.05 | 0.23 | 0.74 | 0.54 | 2.74 | 47 | 9.49 | 183.3 |
| MeanSC | 0.09 | 0.51 | 0.86 | 1.44 | 8.99 | 1.44 | 8.16 | 4.2 |
| EMeanSC | 0.39 | 2.25 | 3.16 | 5.21 | 31.31 | 4.97 | 28.74 | 10.47 |
| MeanShift | 2.91 | 52 | 102 | 67 | 388 | 4138 | 217 | 62388 |
| $k$-means | 0.5 | 36 | 5 | 17 | 1196 | 75 | 302 | 588 |
| ISODATA | 1 | 15 | 5 | 9 | 1178 | 68 | 332 | 337 |
| DBSCAN | 194 | 2731 | 13098 | – | – | 39965 | – | – |
| OPTICS | 638 | 5244 | 40013 | – | – | – | – | – |
| DENCLUE | 6934 | 39849 | – | – | – | – | – | – |

## 4 Conclusion

An experimental comparison of the HCA_MS, HECA_MS, ECCA, CCAE, MeanSC, and EMeanSC algorithms, developed at the FRC ICT, with the nonparametric algorithms DBSCAN, OPTICS and DENCLUE, as well as the *k*-means and ISODATA algorithms from the ENVI software package, has been performed. It was shown that the developed algorithms are superior to the most popular clustering algorithms in terms of clustering quality and/or processing time. In addition, the *k*-means and ISODATA algorithms which are included in common software packages and, as a consequence, are often used in practice, do not allow interactive processing of large multispectral images. On the other hand, the algorithms HCA_MS, HECA_MS, ECCA, CCAE, MeanSC and EMeanSC, thanks to the use of modern approaches to clustering (density-, grid- and ensemble-based), allow dialog-mode image segmentation.

This research was supported by RFBR (grants No. 18-07-00600, 19-29-01175).

## References

1. L. Shapiro, G. Stockman. *Computer vision.* (New Jersey: Prentice Hall, 2001)
2. G. Menardi, Stat. Anal. Data Min.: The ASA Data Sci. J. **13 (1)**. 83-97 (2020)
3. I.A. Pestunov, Yu.N. Sinyavskiy. Bulletin of KemSU. **4/2 (52)**. 110-125 (2012) (In Russ.)
4. D. Krstinic, A.K. Skelin, I. Slapnicar. IET Image Process. **5 (1)**. 63-72 (2011)
5. http://www.harrisgeospatial.com/SoftwareTechnology/ENVI.aspx
6. https://elki-project.github.io
7. https://haifengl.github.com/smile
8. I.A. Pestunov, S.A. Rylov, V.B. Berikov. Optoelectronics, Instrumentation and Data Processing. **51(4).** 329-338 (2015)
9. S.A. Rylov. CEUR Workshop Proceedings. **2033.** 150-155 (2017)
10. S.A. Rylov, I.A. Pestunov. Modern Problems of Earth Remote Sensing from Space. 52 (2017) (In Russ.)
11. I.A. Pestunov, S.A. Rylov, Yu.N. Sinyavskiy, V.B. Berikov. CEUR Workshop Proceedings. **1901.** 194-200 (2017)
12. I.A. Pestunov, Yu.N. Sinyavskiy. Avtometriya. Second IASTED Intern. Multi-Conf. on Automation, Control, and Information Technology – Signal and Image Processing. 5-9 (2005)
13. I.A. Pestunov, V.B. Berikov, Yu.N. Sinyavskiy. Vestnik SibSAU. **5 (31)**. 56-64 (2010) (In Russ.)
14. M. Ester, H.-P. Kriegel, J. Sander, X. Xu. KDD'96. 226-231 (1996)
15. M. Ankerst, M.M. Breunig, H.-P. Kriegel, J. Sander. ACM SIGMOD. 49-60 (1999)
16. A. Hinneburg, D. Keim. Knowl. Inf. Syst. **5 (4)**. 387-415 (2003)
17. Y. Cheng. IEEE Trans. Pattern Anal. **17 (8)**. 790-799 (1995)
18. R.C. Gonzalez, R.E. Woods. *Digital image processing. 2nd ed.* (Addison-Wesley Pub (Sd), 2002)
19. https://drive.google.com/drive/folders/11r-w_RBC38KEYqFJueQ7tlWh6VndDPUn

# Irrigation cooling effect: opportunities in task of estimation of international irrigation water usage in transboundary River Syrdarya basin, Central Asia

*Alexey* Terekhov[1,2*], *Nurlan* Abayev[2,3]

[1] Institute of information and computing technology MES, Almaty, 050010, Kazakhstan
[2] RSE Kazhydromet, Almaty, 050022, Kazakhstan
[3] al-Farabi Kazakh National university, Almaty, 050040, Kazakhstan

**Abstract.** The aim of this research was the analysis of long-term and seasonal dynamics of Irrigation Cooling Effect (ICE) of irrigated arable lands in test sites of Kazakhstan and Uzbekistan in agriculture province "Golodnaya stepp", located in transboundary River Syrdarya basin. The Land Surface Temperature (LST) FEWS NET (Famine early Warning System Network) product was used for calculation and monitoring (2002-2019) ICE values. The amount and efficiency of irrigation water usage significantly affects the ICE values. Therefore, long-term (2002-2019) ICE monitoring on arable land Kazakhstan and Uzbekistan in "Golodnaya stepp" is important as an objective characteristic of irrigation parameters and their changes during the observation period. Analysis of data from 2002-2019 showed that in the key period (May-June) Uzbekistan's arable land is in a better position. Review of two eras 2003-2010 and 2011-2019 it showed that changes in ICE values are directed at: an increase in May, approximately 2.5° in Uzbekistan and 1.3° in Kazakhstan; an increase in July, 1.3° in Uzbekistan and 2.5° degrees in Kazakhstan; and a decrease in September, approximately 1.5° in Kazakhstan, with stable values in Uzbekistan. Thus, ICE monitoring in arid climates is a useful tool for diagnosing water consumption on arable land in various countries, which is especially important in transboundary river basins.

## 1 Introduction

Central Asia region is located at the center of Eurasia, Fig.1. The distance from the oceans leads to significant variability in the seasonal precipitation in mountain systems in the center of Eurasia. Mountain systems of Eurasia, such as the Tien Shan, Pamir, etc. they are zones of river outflow formation of major rivers in Central Asia, including the Syrdarya river, which is the second rang in outflow volume. All the largest rivers in the region have

* Corresponding author: aterekhov1@yandex.ru

ice and snow-ice nutrition and significant seasonal (2-3 times) variability in the annual outflow [1,2].

In the context of the water-scarcity climate of Central Asia, the distribution of transboundary river resources between the countries is an extremely important political issue. Agricultural production in arid climates depends entirely on irrigation. Thus, the food security of Central Asian countries that depend on irrigation agriculture is intricately linked to international agreements on the division of transboundary river resources and their actual implementation. The annual variability in the river outflow makes it difficult to create a transparent water allocation system for cross-border river outflow. Countries in the lower parts of transboundary basins have deficit objective tools to control the actual water use of countries in the upper parts of the basin.

Arable irrigation consumes the bulk of Central Asia's surface runoff. The issue of controlling agricultural water consumption on irrigated arable land according to satellite estimates is still far from being resolved. There are possibilities to control the size of irrigated arable land and assess the state of crops, for example, using vegetation indices [3]. But this is not enough to provide an objective quantitative description of transboundary water allocation. In this paper, we consider Land Surface Temperature (LST) [4] as another satellite parameter that characterizes water usage on irrigated arable land [5,6]. This research uses the Irrigation Cooling Effect (ICE), calculated on the basis of LST, which is the effect of cooling the surface of irrigated arable land, which occurs due to the evapotranspiration of water from the soil surface and vegetation cover, as a result of irrigation.

## 2 Study area

The 2nd largest transboundary river of Central Asia – Syrdarya pass through the territory of Kyrgyzstan, Tajikistan, Uzbekistan and Kazakhstan. River length of more than 2 thousand km, with an average annual flow of about 38 km$^3$ (variation from 19 to 72 km$^3$, maximum seasonal flow in July). At the same time, Kyrgyzstan and Tajikistan are in the upper reaches of the river basin, in the mountainous part of its basin. Uzbekistan occupies the middle part, while Kazakhstan is in the lower part of river basin. Irrigation agriculture is the basis of agricultural production in Uzbekistan. Kazakhstan has the main agricultural lands in the North part (more than 10 million hectares of non-irrigated arable land), and the irrigated arable land in the South of the Republic is important only in the regional aspect, Fig. 1.

For a comparative analysis of the ICE effect of irrigation arable land in Uzbekistan and Kazakhstan, the "Golodnaya stepp" agriculture province was used. A large agricultural province of irrigated arable land with an area of about 800 thousand hectares called the "Golodnaya stepp" is located on the left bank of the Syrdarya river. Most of it belongs to the territory of Uzbekistan (Syrdarya oblast) and a smaller part to Kazakhstan (Makhtaaral district of South Kazakhstan oblast), Fig.2.

Three test sites with a size of 24x18 km were identified for the study. Two polygons in the agricultural zone, on the territory of Kazakhstan (coordinates of the center- 40° 47'N; 68° 20' E) and on the territory of Uzbekistan (40° 37' N; 68° 00' E). As well as a polygon in the natural landscape zone (41° 20' N; 67° 00' E) located on the territory of Kazakhstan, see Fig. 2B

## 3 Materials and methods

FEWS NET products presented for the Central Asia region were used. The Early Warning Explorer (EWX) tool allows users to access the LST Collection 6 products, with a resolution of 3 minutes, which is approximately 5.5 x 4.3 km in the analyzed region. Ten-day renew composites for the period May-September (15 coverages per year) described the seasonal temperature regime of the test sites. The data archive contains information from the period 2002-2020, which made it possible to analyze the eighteen-year period.

ICE value this is LST difference between the natural landscape and irrigated arable land. The decrease in the surface temperature of arable land is a consequence of the irrigation [3]. Irrigation water evaporates, both directly from the capillaries of the soil layer, and through vegetation, and thus the temperature of the underlying surface decreases. The amount of temperature decrease primarily depends on the amount of water. The irrigation type, and cultivated crop varieties and the specifics of agricultural practices also play a role. In case if all other things being equal, the value of ICE depends on the level of water consumption on irrigated arable land.

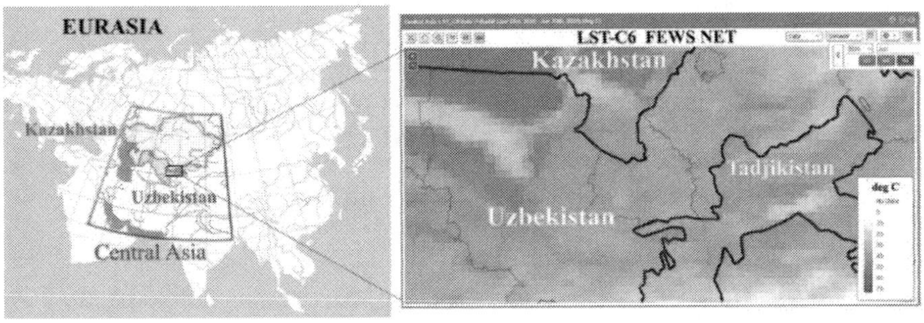


Fig. 1. Disposition of Central Asian region (FEWS NET) and Kazakhstan on Eurasia continent and example of decade "Land Surface Temperature" map of study area from FEWS NET.

Long-term monitoring, which in this research is a period of 18 years, makes it possible to determine the direction of the trend of ICE changes in the period from soil preparation for sowing in May to after harvesting in September. Of particular interest is the time of maximum vegetation, which is usually observed in the second half of July, since it is most closely related to crop yields. Trends in increasing ICE values indicate progressive development of agricultural land use, in which an increasing amount of water is brought to the surface and root layer of the soil.

Of particular interest are comparisons of long-term ICE regimes of irrigated arable land in different parts of the agricultural province [6], related to different countries. In this case, it is arable land of Uzbekistan or Kazakhstan. Access to irrigation water in transboundary river basins usually depends significantly on geographical location. The countries in the lower parts of the basin are most vulnerable. As part of this research, this is Kazakhstan. Long-term comparisons of differences in ICE values between the arable land of Uzbekistan and Kazakhstan characterize the actual water allocation and efficiency of agricultural water usage.

https://doi.org/10.1051/e3sconf/202022302009

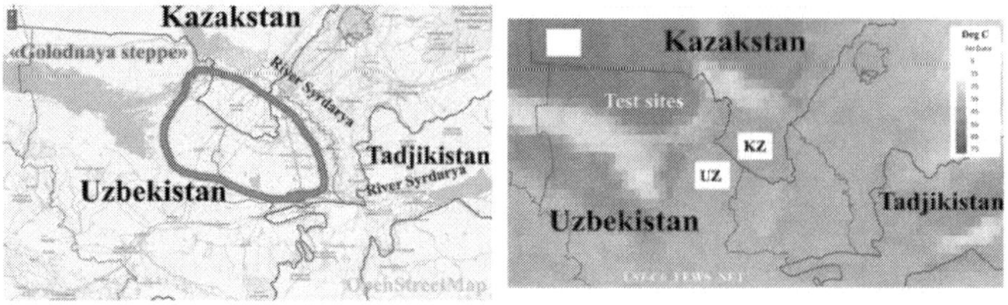


Fig. 2. Regional physical map with "Golodnaya steppe" agriculture province in River Syrdariya basin and test site disposition on fragment of LST FEWS NET map for Central Asian region (3 decade of June 2020).

The data processing technique was extremely simple. Initially, the average values of LST C6 FEWS NET (decadal composites) were determined for three test sites. Thus, time series of ICE values for arable land in Uzbekistan and Kazakhstan were formed. Satellite monitoring included the period from July 2002 to September 2019. Unfortunately, on May 1, 2020, the Sardobinsky reservoir collapsed [7,8]. The water supply regime of the analyzed arable land was distorted by this emergency, which made it impossible to include the 2020 data in the analyzed time series. The time series processing technique itself was based on a linear regression analysis of ICE data series for arable land in Uzbekistan and Kazakhstan, which allowed us to quantify the long-term dynamics of relative water use efficiency in the agricultural sector and national differences in this parameter.

## 4 Results

The calculated ICE values ranged from 5 to 21 degrees. The highest ICE values were recorded in the second decade of August, which is associated with a more intense warming of natural landscapes at this time. The monitoring results are showed in Fig.3-4.

## 5 Discussion

The average long-term (2003-2019) seasonal profiles of ICE values for arable land in Kazakhstan and Uzbekistan in "Golodnaya steppe" agriculture province is shown in Fig. 3. In the period from May to mid-July, the ICE values for arable land in Uzbekistan are 3-4 degrees higher than in Kazakhstan. It seems that this is a consequence of the favorable geographical position of Uzbekistan, located upstream of Kazakhstan along the transboundary River Syrdarya. Uzbekistan provides the best water supply for its arable land during the key period (May-June) of the vegetation season. In July, there is a seasonal maximum water consumption in the River Syrdarya river, which makes it possible to equalize the water needs of irrigation arable land for both countries. In the late summer and early autumn (August-September), the ICE values for Kazakhstan's arable land, on the contrary, exceed the Uzbek values. At the end of the growing season, after harvesting some of crops, water requirements for arable land in Uzbekistan decrease, which automatically reduces the ICE values. Thus, the better water availability of irrigated arable land in Uzbekistan shifts the beginning of the growing season to an earlier date in comparison with Kazakhstan, see Fig. 4

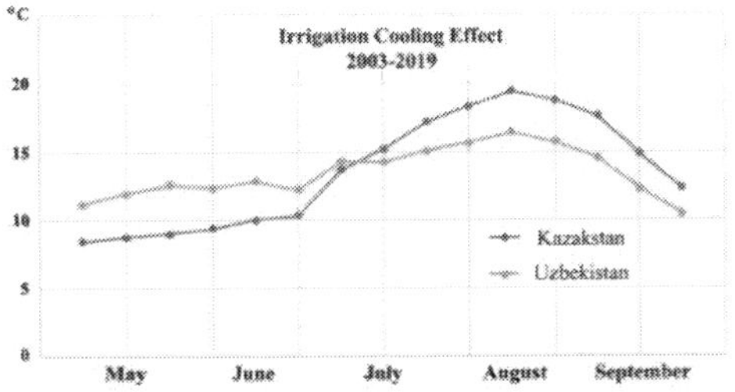


Fig. 3. Long-term average ICE seasonal profiles for test sites in irrigated arable land of Kazakhstan and Uzbekistan in "Gologaya steppe" agriculture province in River Syrdarya basin.

Figure 4a shows a long-term (2002-2019) profile illustrating changes in the ICE values of irrigated arable land in Kazakhstan and Uzbekistan in "Golodnaya steppe" agriculture province for mid-July. The Kazakhstan segment of the province's irrigated arable land shows steady development with an increase in ICE values from approximately 11.5° in 2002-2004 to 17.5° in 2017-2019. There are two abnormally high ICE values in 2005 and 2006, the cause of which is not clear.

Irrigation arable land in Uzbekistan has a more complex evolution of ICE values. There is a ten-year (2007-2016) stability period characterized by an ICE value of approximately 13.5°. The last years of 2017-2019 with high ICE values, about 17°, coincide with the start of operation of the new Sardoba reservoir with a volume of about 1 $km^3$. This reservoir intercepted winter water passes from the Toktogul reservoir (Kyrgyzstan), which operates in an energy mode. May 1, 2020 the Sardobia reservoir was destroyed as a result of the dam breakout [7,8]. There was also a three-year period 2004-2006 with high volatility of ICE values (from 10 to 16 degrees), the reason for which is not known.

Figure 4b shows changes in seasonal profiles of ICE values between two epochs: 2003-2010 and 2011-2019. The difference between these values shows that there is an increase in ICE values in May, both in Uzbekistan and Kazakhstan. However, it is more significant in Uzbekistan: 2.4°C contrary to 1.3°C in Kazakhstan. Also, for arable land in Kazakhstan, a significant increase in ICE values is registered in July, about 2° C, and a decrease in September. The characteristics of Uzbekistan's arable land in the summer-autumn period are generally stable, with some growth in July. The ICE values variations (June-September) almost do not exceed the level of (± 0.5° C).

## 3 Conclusion

The satellite product "Land Surface Temperature" collection 6 FEWS NET with a decade renew, a spatial resolution of 3′ (5,5 x 4, 3 km) and an archive from 2002 it allows to calculate and monitor Irrigation Cooling Effect values for irrigation arable land in countries located in arid climates. The dynamics of long-term changes of ICE values provides a basis for evaluating the effectiveness of irrigation water usage in Kazakhstan and Uzbekistan in the "Golodnaya steppe" agriculture province (Central Asia). There appears the possibility to diagnose the accessibility of transboundary water resources of the River Syrdarya for irrigation arable land of different countries and changes in the long-term aspect. Such estimates are most interesting for countries in the lower parts of transboundary river basins, since there is a lack of objective tools for controlling irrigation water usage in the upper parts of the basin, because the natural significant variability of river outflow in arid climates makes the situation significantly uncertain.

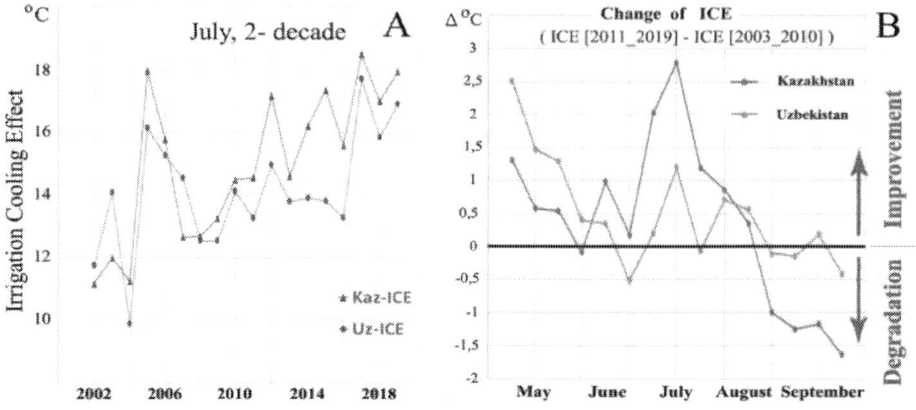


Fig. 4. A - Long-term evaluation of July's ICE values of national test sites in the agricultural province "Golodnaya steppe" in transboundary River Syrdarya basin: B - Modern directions of seasonal ICE value change in national test sites in the agricultural province "Golodnaya steppe" in transboundary River Syrdarya basin.

## Acknowledgment

This research has been funded by the Science Committee of the Ministry of Education and Science of the Republic of Kazakhstan (Grant No. AP08957394).

## References

1. A. Terekhov, I.S. Vitkovskaya, N. Abayev, K., E3S Web of Conference, **149**, 03004, (2020), doi: 10.1051/e3sconf/202014903004.
2. A. Terekhov, N. Abayev, I. Vitkovskaya, A. Pak, Z. Egemberdieva, Sovremennye problemy distantsionnogo zondirovaniya Zemli iz kosmosa, **17**, 2, 275-281, (2020), doi: 10.21046/2070-7401-2020-17-2-275-281.
3. A. Terekhov, N. Abayev, E. Lagutin, Sovremennye problemy distantsionnogo zondirovaniya Zemli iz kosmosa, **17**, 1, (2020), doi: 10.21046/2070-7401-2020-17-1-128-138.
4. A. Barnston, P. Schickedanz, J. Climate Appl. Meteor., **23**, 865-888, (1984), doi: 10.1175/1520-0450(1984)023<0865:TEOIOW>2.0.CO;2
5. S. Han, Zh. Yang, Environ. Res. Lett., **8**, 2, 024039, (2013), doi: 10.1088/1748-9326/8/2/024039.
6. Q. Yang, X. Huang, Q. Tang, Sci. Total Environ., **705**, 135984, (2020), doi: 10.1016/j.scitotenv.2019.135984.
7. A. Terekhov, N. Abayev, E. Lagutin, Sovremennye problemy distantsionnogo zondirovaniya Zemli iz kosmosa, **17**, 3, (2020), doi: 10.21046/2070-7401-2020-17-3-255-260.
8. A. Konstantinova, E. Loupian, Sovremennye problemy distantsionnogo zondirovaniya Zemli iz kosmosa, 2020, **17**, 3, (2020), doi: 10.21046/2070-7401-2020-17-3-261-266.

# Application of high frequency filtration to remove cloudiness in Earth remote sensing images

*Valeriy* Tutatchikov[1*], *Mikhail* Noskov [1]

[1] Siberian Federal University, School of Space and Information Technology, 660074, Krasnoyarsk, Russian Federation

**Abstract.** At present, methods of digital processing of Earth remote sensing images are widely used to improve the image quality. For example, many images are discarded due to high clouds in the images, which obscure objects of interest. In this paper, the possibility of using high-frequency global filters to reduce cloudiness in the image is considered, and the results of image enhancement are shown.

## 1 Introduction

At present, methods of digital processing of Earth remote sensing (ERS) images are widely used to improve the image quality. But many original images contain distortions, for example, high clouds, which partially or completely hide objects of interest. The possibility of preliminary ranking or selection of images with respect to the prevalence and location of clouds in the image should significantly reduce the search time for suitable images for solving applied problems [1].

Another common way to improve such images is preliminary processing of remote sensing data: radiometric, atmospheric or geometric correction [2]. It is also possible to use classical methods of spatial filtering of digital image processing [3]. In this paper, we will consider the possibility of applying global high-frequency filtering methods to eliminate cloudiness of remote sensing images.

## 2 Method description

Consider, as the initial data, a snapshot of remote sensing of the Earth of the vicinity of the Taseevsky district of the Krasnoyarsk Territory (Figure 1). As a distortion of the cloudiness of the image, the smoke in this area is presented, and the central part of the image contains a high degree of distortion, and the lateral left one is already less [4].

* Corresponding author: vtutatchikov@mail.ru

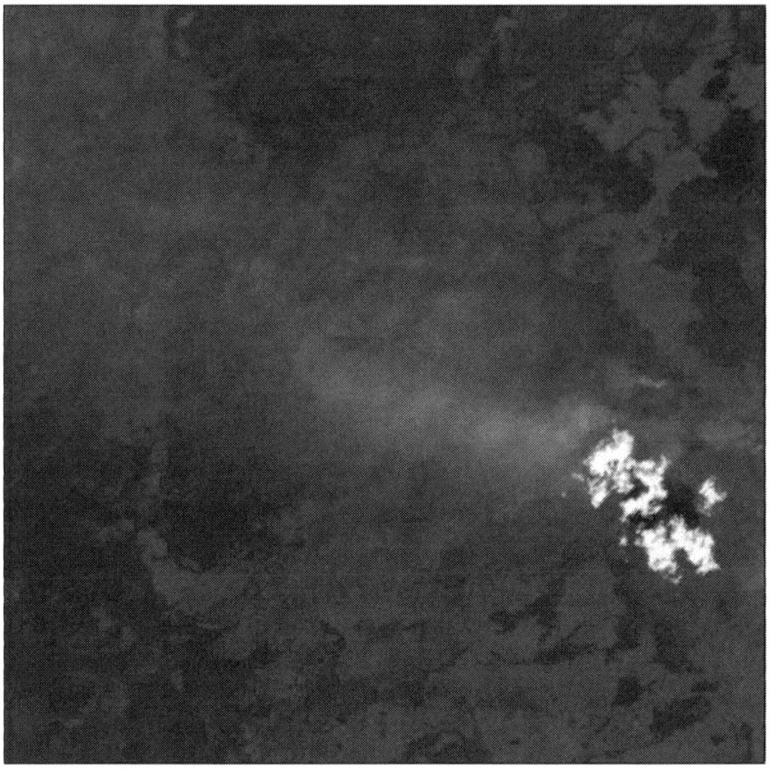

**Fig. 1.** Surface snapshot with smoke

For the convenience of calculations, let us set the size of the investigated area as a power of two: 512 * 512 pixels (the use of the Fourier transform of two-dimensional signals with a number of samples other than a power of two is described in [5-6]). This image can be represented as a function *f(x,y)* of the brightness of a pixel with coordinates *x* and *y*, taking integer values 0-255 for each red, green, blue channel in the RGB color model.

In [3], it is proposed to multiply all the image elements *f(x,y)* by $(-1)^{x+y}$, and then calculate the discrete Fourier transform of this function. In this case, we get a centered Fourier transform *F(u,v)* in the frequency domain, in which high frequencies (responsible for image contours and sharp transitions) will be in the center of the spectrum, and low frequencies (responsible for image smoothing and color saturation) will be around the edges.

We apply a global high-pass filter to the obtained function *F(u,v)*, for example, a Gaussian high-pass filter (HPF):

$$H(u,v) = 1 - \exp\left(\frac{-D^2(u,v)}{2 \cdot D_0^2}\right)$$

where $D(u,v)$ is the distance from the center of the Fourier transform to the point with coordinates $(u,v)$, $D_0$ is the cutoff frequency. The Gaussian HPF graph is shown in Figure 2.

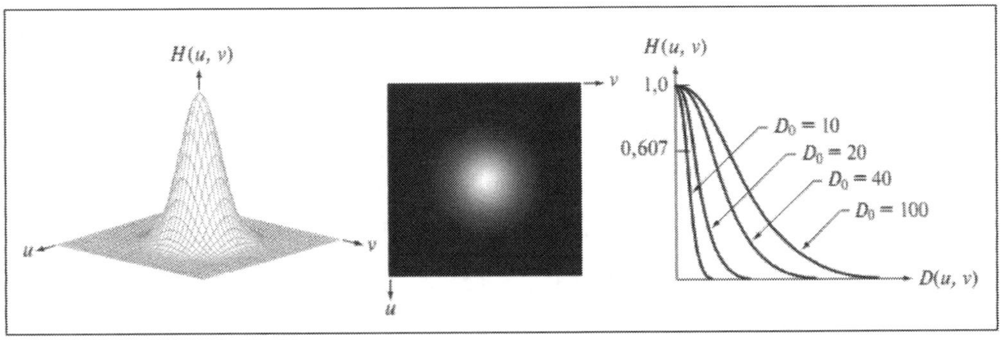


**Fig. 2.** Graph of a Gaussian LPF function on the left, a grayscale image of the filter in the center, radial filter profiles for different values of $D_0$ on the right

As a result of filtering, the high frequencies will remain unchanged, and the low frequencies outside the radius $D_0$ will be zeroed. After that, we calculate the two-dimensional inverse Fourier transform, multiply all samples of the resulting function by $(-1)^{x+y}$ to cancel centering, and get the processed image $f_1(x,y)$.

## 3 Obtained results

The result of filtering is shown in Figure 3. The image has sharp outlines of objects, there is no color saturation. At the same time, the rather rare smoke on the left side of the image is no longer visible, but a denser, outlined haze in the center remains noticeable.

**Fig. 3.** The result of applying a Gaussian HPF with $D_0 = 4$

Now let's combine the obtained filtering result with the original image. The final result of digital processing is shown in Figure 4.

**Fig. 4.** Global high pass filtering result

## 4 Conclusion

On the example of a photo of a smoky surface of the Taseevsky district of the Krasnoyarsk Territory, a method of applying global high-frequency filtering to improve the image is shown. As a result of processing, the sparse haze on the left side of the original image becomes less noticeable. At the same time, dense smoke in the center became more rarefied, but did not disappear completely, since in the high-frequency spectrum the boundaries of this cloudiness have a significant effect on the image.

## Acknowledgments

This work was carried out as part of the state assignment of the Ministry of Science and Higher Education of the Russian Federation (scientific code FSRZ-2020-0011).

## References

1. E.A. Maltsev, E.E. Sirotin, D.A. Perfil'ev, G.M. Tsibul'skii, *Journal of Siberian Federal University. Engineering & Technologies 2*, **№5**, 229 – 242 (2012)

2. O.S. Tokareva, *Processing and interpretation of Earth remote sensing data*, 148 (2010)
3. R.C. Gonzalez and R.E. Woods. *Digital Image Processing,* (2012)
4. Snapshot of the Taseevsky district of the Krasnoyarsk Territory from 07.30.2020. Access mode: http://earthexplorer.usgs.gov/
5. M.V. Noskov, V.S. Tutatchikov, *Pattern Recognition and Image Analysis*, **25**(1), 81-83 (2015)
6. M. Noskov, V. Tutatchikov, *E3S Web of Conferences*, **149**, 02010 (2020)

# Computer model of high-precision navigation of a small satellite constellation for remote Earth sensing tasks

*Valeriy* Tutatchikov[1*], *Oleg* Drozd[1] и *Denis* Kapulin[1]

[1] Siberian Federal University, School of Space and Information Technology, 660074, Krasnoyarsk, Russian Federation

**Abstract.** One of the key tasks in the formation of the small satellite constellation is to ensure its stable configuration in time. The specific of this problem lies in the fact that the task must be solved in real time with sufficiently high accuracy with significant restrictions on the power consumption of onboard satellite systems. The paper briefly discusses one possible approach of secondary signal processing and solving the navigation and ballistic tasks, which involves the joint use of code, phase measurement methods, integer phase ambiguity resolution, and inter-satellite measurements.

## 1 Introduction

It can be argued that by now conditions have emerged for the creation and deployment of fully functional space systems based on orbital constellations of of the small satellite (SST), which are considered not as an alternative to existing systems based on medium and heavy spacecraft, but as an addition, significantly expanding the solution applied tasks in various fields of activity.

SST open up wide opportunities for the synthesis of space systems with new qualitative capabilities. For example, for space systems of Earth remote sensing (ERS), the following systemic effects are possible:

1. Multi-position, multi-range and simultaneous observation of an object from several SSTs, providing the effect of stereo photography.

2. Possibility of integrating information received from heterogeneous onboard complexes of special equipment of SST, leading to an increase in the information content of observation.

3. Ensuring the continuity of observation due to the possibility of "passing the baton" of observation to other SST of the orbital constellation that are sequentially entering the object's visibility zone.

The use of SST for remote sensing purposes has great prospects and can significantly improve the consumer qualities of space information, but at the same time it requires solving a number of scientific and technical problems. Along with the system-wide tasks of

* Corresponding author: vtutatchikov@mail.ru

maintaining the integrity of the orbital constellation, it is required to solve the tasks of autonomous navigation-ballistic and frequency-time support [1].

At present, for autonomous navigation support of autonomous small satellite, onboard navigation systems based on optical-electronic sensors of celestial bodies (Earth, Sun, navigation stars), inertial attitude control systems, multichannel receivers of signals from global navigation satellite systems (GNSS) are used [3–5]. For ground support of small satellite operation, networks of ground posts for radar and optical control of the current satellite position are used [6–9]. One of the key disadvantages of the existing control facilities is their dependence on the ground segment of the group control and the presence of gaps in the navigation field associated with the territorial location of control posts. In this paper, we propose a solution to the problem of high-precision navigation and ballistic support of a large group of SST in autonomous mode without the use of ground facilities and with minimization of the cost of the proposed solutions [2].

## 2 Method description

The proposed methodology for the navigation support of the SST constellation is a new combination of known methods of secondary processing of navigation information received from GNSS GPS and GLONASS. Solving the navigation problem by the onboard means of the small spacecraft in autonomous mode includes the following procedures:

1) individual and group calculation of the absolute and relative coordinates of the SST by means of code and phase measurements [10, 11];

2) calculation of the motion parameters of the center of mass of the small satellite (components of the velocity vector) based on the results of the assessment of pseudo-velocities in the code and phase modes;

3) solution of the integer ambiguity of phase measurements of pseudo-range by methods based on the use of second phase differences, calculated using the first phase differences [12, 13];

4) provision of mutual navigation measurements by means of onboard equipment for inter-satellite measurements and inter-satellite communication lines [14, 15].

## 3 Experimental results

To test the theoretical principles of the implementation of these procedures as applied to a constellation of small satellite with a minimum composition: the master (leader) and the slave satellites, a computer model of navigation support was developed. The simulation model is implemented in C ++ in the Qt development environment. The results of calculating the absolute coordinates and speed of the master small spacecraft are shown in Figure 1, relative coordinates and speed of the slave small spacecraft - in figure 2.

The simulation results showed that the simulation model of the autonomous navigation support of the small satellite constellation meets the requirements for solving the navigation-ballistic problem, in particular, in terms of the root-mean-square error in measuring relative coordinates - no more than 0.02 m and relative speed - no more than 0.01 m / s (figure 3).

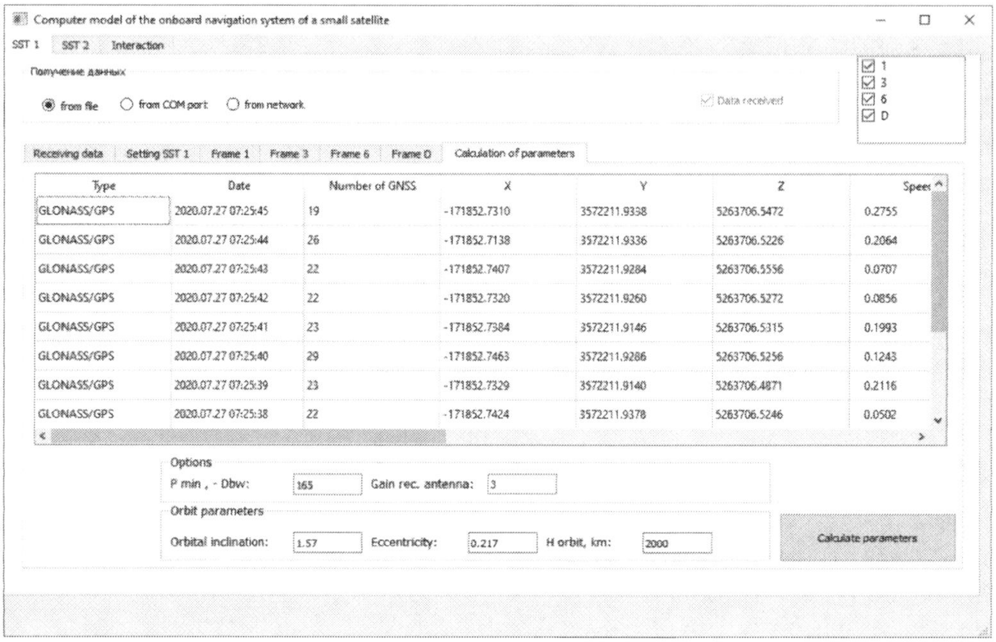


**Fig. 1.** The absolute coordinates and speed of the leading SST

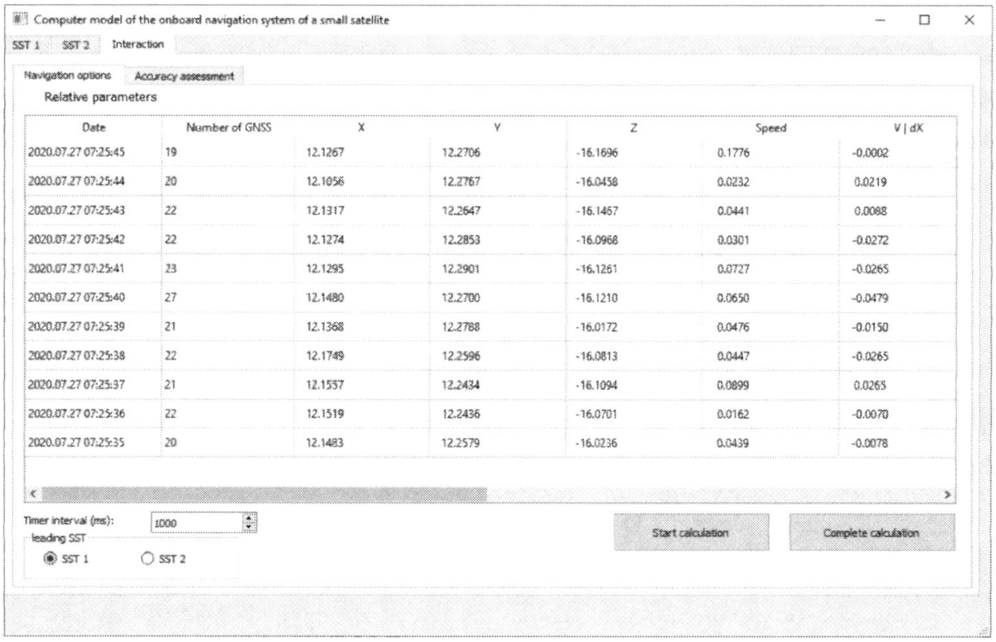


**Fig. 2.** Relative coordinates and speed of the slave SST

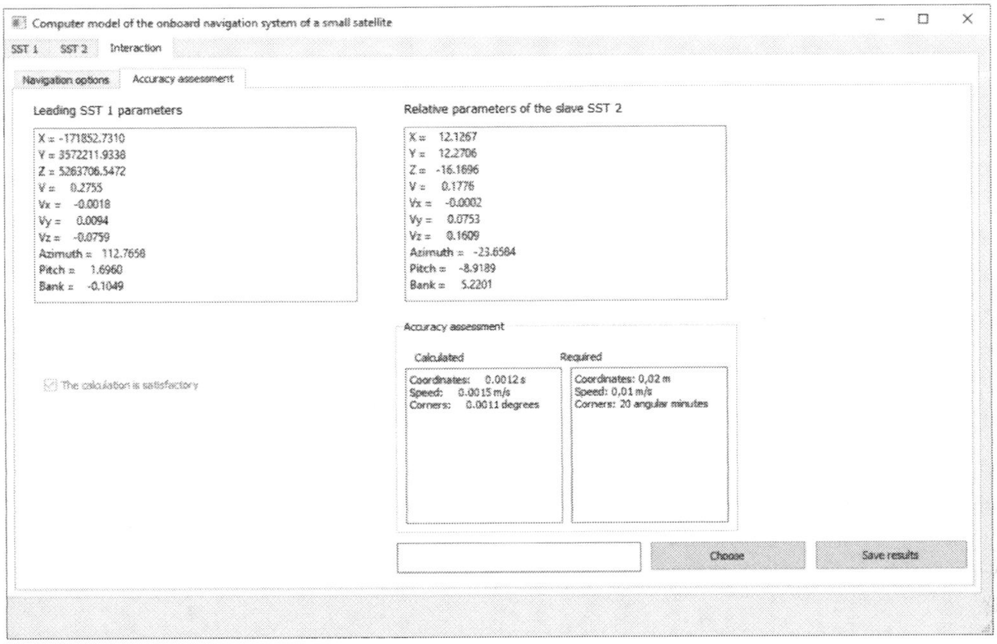


**Fig. 3.** Estimation of the accuracy of calculating the coordinates and speed of the slave SST

## 4 Conclusion

The development of methods and means of navigation support for high-precision autonomous navigation of constellation of small satellite for performing the tasks of remote sensing of the Earth involves the integration of various methods of processing navigation signals. The paper briefly considers one of the options for such a complex procedure for the secondary processing of navigation signals and the solution of the navigation-ballistic problem, involving the combined use of methods of code, phase measurements, solution of integer phase ambiguity, and intersatellite measurements.

## Acknowledgments

The work was supported by the Ministry of Education and Science of the Russian Federation (Agreement № 075-15-2019-1684, unique ID project RFMEFI60819X0274).

## References

1. J.R. Kopacz, R. Herschitz and J . Roney, *Acta Astronaut,* **170**, 93–105 (2020)
2. X. Ma, J. Fang and X. Ning, *Prog. Aerosp. Sci.*, **63**, 56–66 (2013)
3. X. Sun, C. Han and P. Chen, *Aerosp. Sci. Technol.*, **67**, 228–36 (2017)
4. D.D. Dmitriev, V.N. Tyapkin, Y.L. Fateev, A.B. Gladyshev and P.Y. Zverev, *Moscow Workshop on Electronic and Networking Technologies (MWENT)*, 1–5 (2020)
5. G. Giorgi, T.D. Schmidt, C. Trainotti, R. Mata-Calvo, C. Fuchs, M.M. Hoque, J. Berdermann, J. Furthner, C. Günther, T. Schuldt, J. Sanjuan, M. Gohlke, M. Oswald, C. Braxmaier, K. Balidakis, G. Dick, F. Flechtner, M. Ge, S. Glaser, R. König, G.

Michalak, M. Murböck, M. Semmling and H. Schuh, *Adv. Sp. Res.*, **64**, 1256–73 (2019)

6. S. Sgubini and G.B. Palmerini, *IEEE Aerospace Conference*, 1–7 (2010)
7. Y. Liu, Y. Chen, Y. Jiao, H. Ma and T. Wu, *IEEE Access*, **8**, 63923–34 (2020)
8. K. Riesing, H. Yoon and K. Cahoy, *IEEE International Conference on Space Optical Systems and Applications (ICSOS)*, 108–14 (2017)
9. D. Strugarek, K. Sośnica and A. Jäggi, *Adv. Sp. Res.*, **63**, 417–31 (2019)
10. P. Teunissen and O. Montenbruck, *Springer International Publishing* (2017)
11. A. Oxley, *A Mathematical Discourse ed Academic Press* (2017)
12. A. Raskaliyev, S.H. Patel, T.M. Sobh and A. Ibrayev, *IEEE Access*, **8**, 24873–86 (2020)
13. D.D. Dmitriev, Y.L. Fateev and V.N. Ratushniak, *International Siberian Conference on Control and Communications (SIBCON)*, 1–4 (2019)
14. A.M. Crisan, A. Martian, R. Cacoveanu and D. Coltuc, *Measurement*, **154**, 107479 (2020)
15. F. Yu, Z. He and N. Xu, *Acta Astronaut.*, **160**, 646–55, (2019)

# A New Nonparametric Algorithm for Preprocessing Stochastic Data with Uncertainty

*Ekaterina* Chzhan[1*]

[1]Siberian Federal University, School of Information and Space and Technology, 660074 Krasnoyarsk, Russian Federation

**Abstract.** The article deals with the problem of modeling stochastic processes under uncertainty. The peculiarity of the processes under consideration is that the researcher does not have information about the mathematical structure of the object; the object is represented as a black box. The article proposes to use a nonparametric modeling algorithm based on a nonparametric estimate of the regression function on observations. To improve the accuracy of modeling, it is proposed to use an algorithm for generating training samples. The algorithm differs from the previous modification by the definition of essential variables. The results of computational experiments have shown the effectiveness of the proposed algorithms.

## 1 Introduction

The problem of data analysis and modeling of stochastic processes is relevant for many subject areas, including in the case of processing geoinformation data [1]. In the case when the mathematical structure of the object under study is unknown, one of the solutions can be the use of nonparametric estimates [2]. As it is known, the accuracy of nonparametric models depends on many factors. One of the key factors is the quality of the initial data [3]. In the original data there may be outliers due to measurement errors and missing data due to different discreteness of control [4]. These features can negatively affect the quality of modeling, and in some cases lead to unsatisfactory results. Another feature is the uneven sparse distribution of the sample elements in the space of input and output variables [5]. In this case, resampling algorithms [6] or bootstrap methods [7] can be used. If the sample sizes are small with a large space dimension, it is proposed to use methods for generating training samples. However, the previously proposed methods for preprocessing sparse samples cannot be applied in conditions of noisy data (in the presence of outliers), as well as in the presence of a large number of variables, including insignificant ones. Modification of existing algorithms for obtaining training samples will expand the conditions for the applicability of these methods and improve the accuracy of modeling.

The rest of the article is organized as follows. In Section 2, we present the problem statement. Section 3 contains a proposed algorithm our, and in Section 4 we analyze the results of our simulation study data. Discussion and conclusions are presented in Section 5.

* Corresponding author: echzhan@sfu-kras.ru

## 2 A problem statement

Consider the proposed process modeling scheme presented in Figure 1 [8]. The following designations are accepted here: $x(t)$ is an output process variable, $\boldsymbol{u}(t)$ is an input vector action that can be controlled and measured. Random action is $\xi(t)$, control blocks of input and output variables are $G^u$ and $G^x$. Random noises $g^x(t)$, $g^u(t)$ have zero mathematical expectation and limited variance. Measurements of input and output variables form a sample of observations $\{\boldsymbol{u}_i, x_i, i=\overline{1,s}\}$, which is given to a data analysis block. This module contains data preprocessing algorithms for training sample generating.

The modeling task is not only to obtain an estimate $\hat{x}(t)$ of the output process variable $x(t)$, but also to process the initial data.

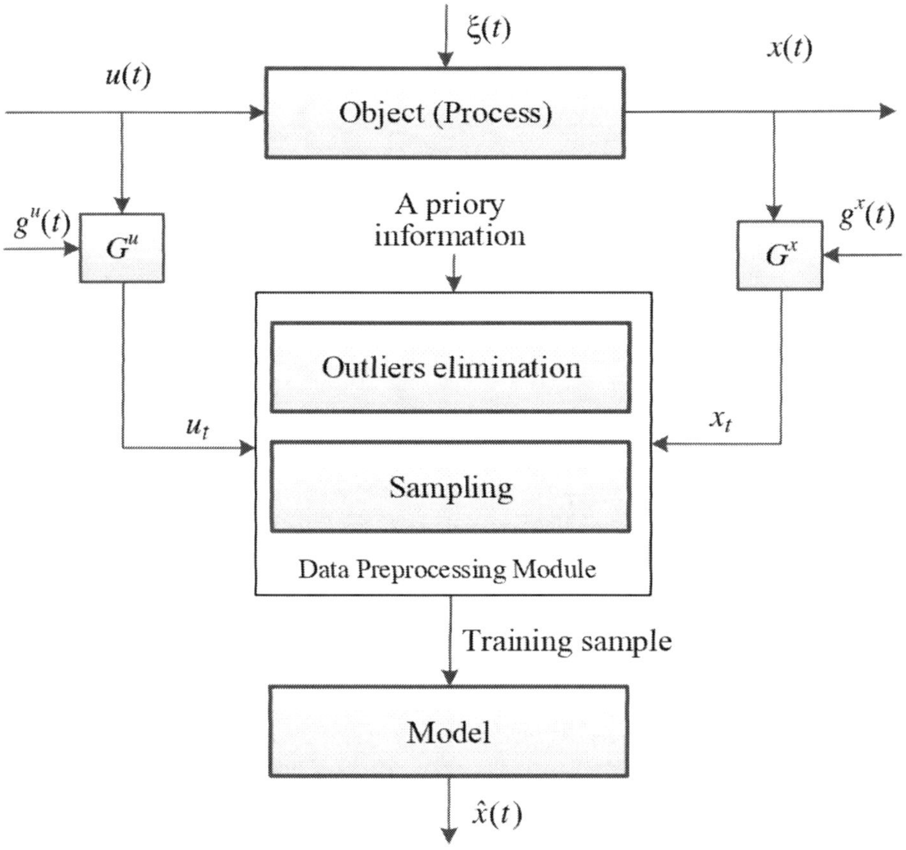


**Fig. 1.** A discrete-continuous process identification scheme.

## 3 A new modification of nonparametric algorithm for data preprocessing

As a model, we use the well-known nonparametric Nadaraya-Watson (NW) estimator [9, 10]:

$$\hat{x}(\boldsymbol{u}) = \frac{\sum_{i=1}^{s} x_i \prod_{v=1}^{m} K\left(\frac{u^v - u_i^v}{c_s^{u[v]}}\right)}{\sum_{i=1}^{s} \prod_{v=1}^{m} K\left(\frac{u^v - u_i^v}{c_s^{u[v]}}\right)}, \quad (1)$$

where $\boldsymbol{u} = (u_1, u_1, \ldots, u_m)$, $c_s$ is the bandwidth or smoothing parameter, which is unique for every variable, $K\left(\frac{u^v - u_i^v}{c_s^{u[v]}}\right)$ is a kernel function.

The estimate belongs to the class of local approximation, the accuracy depends on the number of observations that are in the $c_s$-neighborhood of the point under consideration. An algorithm for generating additional observations was proposed earlier [11]. However, if there are outliers in the sample, the result will be unsatisfactory. In addition, all variables were previously used to calculate the estimate (1) in [12]. As it is known, the inclusion of insignificant variables in the model can lead to a decrease in accuracy. Therefore, it is proposed to identify significant variables and only use them both for calculating the estimate and in the case of generating a new training sample The paper proposes to modify the algorithm as follows:

1. Outliers identification and treatment using nonparametric method [13].
2. Determination of smoothing parameters $c_s^{u[1]}, c_s^{u[2]}, \ldots, c_s^{u[m]}$ by minimizing the criterion of compliance of the object's output $x(\boldsymbol{u})$ with the model $\hat{x}(\boldsymbol{u})$ in the cross-validation mode:

$$I\left(c_s^{u[1]}, c_s^{u[2]}, \ldots, c_s^{u[m]}\right) = $$
$$= \sum_{l=1}^{s}\left(\hat{x}\left(u_{1,l}, u_{2,l}, \ldots, u_{m,l}\right) - x_i\right)^2 = \min_{c_s^{u[1]}, c_s^{u[2]}, \ldots, c_s^{u[m]}}, l \neq i \quad (2)$$

   where the index $i$ is the observation number in (1). The Nelder-Mead multidimensional optimization method can be used as an optimization algorithm. This method allows you to work with nonsmooth and noisy functions.
3. Among all the found coefficients $c_s^{u[1]}, c_s^{u[2]}, \ldots, c_s^{u[m]}$, select the maximum value. This coefficient corresponds to the least significant component $\boldsymbol{u}$.
4. Calculate $\hat{x}(\boldsymbol{u})$ due to (1) excluding the factor $\Phi(\cdot)$ for which the optimal value of the smoothing parameter is maximum.
5. Calculate the value of the mean squared error (MSE):

$$\text{MSE} = \frac{1}{s}\sum_{i=1}^{s}\left(\hat{x}_i - x_i\right)^2 \quad (3)$$

6. The steps 2 – 5 will be repeated until $\text{MSE}_n \leq \text{MSE}_{n-1}$. Thus, all significant variables will be found. Let there be $k \leq m$ variables.
7. Calculate the average number $\lambda$ of sample elements that are found in the $c_s$ neighborhood of each sample element.
8. In the vicinity of observations $\dot{u}$, for which the number of neighbors is smaller than the sample mean $\lambda$, additional artificial sample elements are generated due to the formula:

$$\tilde{u}_{li}^j = \dot{u}_i^j + \psi_l^j c_s^{u[l]}, j = \overline{1,k}, i = \overline{1,s'}, l = \overline{1,v_i}, \quad (4)$$

where $\psi_i^j \neq 0$ is a random variable distributed according to a uniform law in the interval [-1; 1], $\dot{u}$ is the value of the input variables of the element, in the vicinity of which the values $\tilde{u}_{ii}^j$ are generated.

9. For the generated items, instead of the object output we calculate estimation (1) with only significant components of vector $\boldsymbol{u}$. The generated elements and the original sample constitute a new training sample $\left\{\tilde{\boldsymbol{u}}_i, \tilde{x}_i, i = \overline{1, s_1}\right\}, s_1 \geq s$.

We obtain the Nadaraya-Watson kernel estimator (NWN) with new training sample $\left\{\tilde{\boldsymbol{u}}_i, \tilde{x}_i, i = \overline{1, s_1}\right\}$ as follows:

$$\hat{x}_{NWN}(\boldsymbol{u}) = \frac{\sum_{i=1}^{s_1} \tilde{x}_i \prod_{v=1}^{k} \Phi\left(\frac{u^v - \tilde{u}_i^v}{c_s^{u[v]}}\right)}{\sum_{i=1}^{s_1} \prod_{v=1}^{k} \Phi\left(\frac{u^v - \tilde{u}_i^v}{c_s^{u[v]}}\right)} \quad (5)$$

## 4 Simulation results

A simulation study was conducted to compare the performances of the estimators with the classical Nadaraya-Watson estimators. For the simulation, we used the following function:

$$x(u) = 7u_1 + 6\sqrt{u_2} - 8u_3 + 2u_4^2 + 0.1u_5 + \xi \quad (6)$$

where the $u_l, l = \overline{1,5}$ were drawn from a uniform distribution based on the interval [0, 1]. The $\xi$ have a normal distribution with 0 mean and 0.1 variance. In this way, we generated samples of size 200, 500 and 1000. Samples were sparse with outliers.

As a kernel function we used Epanechnikov kernel function [12]:

$$K(z) = \begin{cases} 3(1 - z^2)/4, \text{if } |z| \leq 1, \\ 0, \text{else} \end{cases} \quad (6)$$

For each sample, we computed the values of mean percentage absolute error (MAPE) related to the kernel estimations: standard Nadaraya-Watson estimation (1) NW, estimation NWS (1) in case of the training sample obtained with method described in [11] and estimation NWN (5). The MAPE values of the kernel estimators which are obtained using the Epanechnikov kernel functions (6) are given in Table 1.

**Table 1.** MAPE values of the estimations

| s | NW | NWS | NWN |
|---|---|---|---|
| 200 | 39.3 | 19.5 | 18.8 |
| 500 | 28.9 | 18.9 | 18.1 |
| 1000 | 24.3 | 18.1 | 17.3 |

As seen from Table 1, for all sample sizes, the kernel estimators NWN in case of usage the training sample obtained with the proposed algorithm of data preprocessing have smaller MAPE values than the NWS kernel estimator. In each case, it is seen that NWS has the best performance.

## 5 Conclusion

In the paper, we have studied the new modification of data preprocessing algorithm. The results of the simulation study, which was performed to evaluate the performances of the kernel estimators considered, showed that the use of training samples obtained with the proposed algorithm has improved the accuracy of modeling.

**Acknowledgments**: the study was supported by a grant from the President of the Russian Federation for state support of young scientists MK-763.2020.9.

## References

1. D. Zhang, Y. Zhang, H. Li. Ad. in Marine Sc. **1**, 013 (2009)
2. A. B. Tsybakov. *Introduction to nonparametric estimation*. Springer Science & Business Media, (2008)
3. B. Busygin, S. Nikulin. *Specialized geoinformation system RAPID: features, structure, tasks* in 14th EAGE International Conference on Geoinformatics-Theoretical and Applied Aspects, **1** (2015)
4. S. Efromovich. *Missing and modified data in nonparametric estimation: with R examples*. CRC Press (2018)
5. J. Chen, D. Zhang, W. Zhou, Z. Chen, H. Li. (2020). Str. and Infrastr. Eng. (2020)
6. Ł. Smaga. Com. in St.-Sim. and Comp. **46**, 10 (2017)
7. M.R. Chernick, W. González-Manteiga, R.M. Crujeiras, E.B. Barrios. *Bootstrap methods* (2011)
8. V. Vapnik. *The nature of statistical learning theory*. Springer Science & Business Media, (2013)
9. W. Hardle. *Applied Nonparametric Regression*, Cambridge, New Rochelle (1990)
10. S. Demir, O. Toktamis. On the adaptive Nadaraya-Watson kernel regression estimators, H. J. of M. and St. **39**, 3 (2010)
11. A.V. Medvedev, E.A. Chzhan. *On nonparametric modelling of multidimensional noninertial systems with delay*, B. of the S. Ur. St. Un. Ser. M. M., Pr. & Com. S. **10**, 2 (2017)
12. M. A. Denisov, E. A. Chzhan, A. A. Korneeva. *Non-parametric approach for preliminary processing of earth remote sensing data* in E3S Web of Conferences, **75**, 01015 (2019)
13. G.S. Watson. *Smooth regression analysis*, In. J. of St., **26**, 4 (1964)

# Information content of statistical texture features in the problem of recognition and mapping of natural and man-made objects from space images

*Egor* Dmitriev[1,2], *Anton* Sokolov[3], *Sergey* Zotov[2], *Timophey* Kondranin[2], *Petr* Melnik[4]

[1]Marchuk Institute of Numerical Mathematics of Russian Academy of Sciences, ul. Gubkina 8, 119333 Moscow, Russia
[2]Moscow Institute of Physics and Technology (National Research University), 9 Institutskiy per., Dolgoprudny, Moscow Region, 141701, Russia
[3]Laboratoire de PhysicoChimie de l'Atmosphère Université du Littoral Cote d'Opale, , Avenue Maurice Schumann 189A, 59140 Dunkerque, France
[4]Federal State Budgetary Educational Institution of Higher Education "Bauman Moscow State Technical University (National Research University) ", 105005, Moscow, 2nd Baumanskaya st. 5, bld. 1.

**Abstract.** Statistical texture features are frequently used for the thematic processing of very high spatial resolution satellite images. The assessment of information content of 1st and 2nd order statistics is carried out based on processing WorldView-2 images of test areas located on the territory of the Savvatyevskoe forestry and employing the corresponding ground-based data. The comparison of the accuracy and computational efficiency of traditional and ensemble classifiers in the problem of pattern recognition of various natural and man-made objects reveals the high performance of the error correcting output codes method. The estimates obtained in this study demonstrate the advantage of using ensemble classification and 2nd order statistical texture features.

## 1 Introduction

An important direction of the development of methods for thematic processing of Earth remote sensing data is the introduction of jointly used spectral and texture characteristics of ground objects. The advantages of spectral-texture classification are shown in different studies. For example, the method for combined processing of multispectral and panchromatic satellite images of high spatial resolution, proposed in the paper [1], shows the effectiveness of this approach in solving the problem of classification of soil and vegetation cover in comparison with traditional approaches based on classification by spectral features [2].

Similar conclusions can be found in other scientific papers. For example, when considering the problem of remote sensing identification of species composition of tropical forests, it was revealed that the combined use of the spectral features of stands and statistical texture analysis of panchromatic images can improve the classification accuracy to acceptable values [3]. At the same time, it was shown that such a result cannot be obtained on the basis of characteristic changes of spectral features in different seasons.

High spatial resolution becomes necessary when assessing changes in structural changes of vegetation cover. Structural parameters of forest areas are primarily determined by forest inventory relative stocking and species composition. Estimates of these parameters obtained from remote sensing data are further used in statistical models for the assessment of the biological productivity of forest stands. A good example is results of processing satellite images of Quickbird and Pléiades of the Land de Gascony natural park, located in the southwestern part of France, presented in [4]. In addition to stand density connected with the relative stocking, the proposed method made it possible to obtain estimates of the size of crowns, height and diameter of trees.

Currently, there are many different methods for extracting texture features, however, the issue of assessing their information content in the problems connected with the thematic processing of aerospace images remains open. In this study we present the results of evaluating the effectiveness of using statistical features of the 1st and 2nd orders in the implementation of texture classification of satellite images of very high spatial resolution based on algorithms of varying degrees of complexity.

## 2 Texture classification method

The statistical approach to the extraction of texture features from panchromatic images implies an assessment of the probability function of the distribution of gray levels for pixels within the moving window and the subsequent calculation of certain statistical parameters [5]. Currently, statistical texture characteristics of 1st and 2nd orders are used for thematic satellite image processing. For the extraction of the 1st order characteristics, we need to estimate the probability mass function for the gray level distribution, i.e. the occurrence of each of the possible values of gray levels in the analyzed image should be found. In this study, we consider 5 statistical characteristics of the 1st order: mean, mean square, entropy, energy and variance. The corresponding calculation formulas are given in [5].

The calculation of the 2nd order characteristics is based on the construction of the probability mass function for the joint occurrence of gray levels (i.e. the normalized gray level co-occurrence matrices, GLCM) for the given parameters of distance and direction of adjacency [6]. For the calculations, we use 19 different characteristics: autocorrelation, cluster prominence, cluster shade, contrast, correlation, difference entropy, difference variance, dissimilarity, energy, entropy, homogeneity, homogeneity2, information measure of correlation 1, information measure of correlation 2, maximum probability, sum average, sum entropy, sum squares, sum variance. The calculation formulas for these parameters are presented in [1].

We used different supervised learning classification methods of varying degrees of complexity for solving the problem of recognition of natural and man-made objects by texture characteristics. The detailed description of these methods can be found in [7] and [8]. Traditional metric classifiers are the Nearest Centroid (NC) and K Nearest Neighbors (KNN) with optimized kd-tree search. Linear (LDA) and quadratic (QDA) discriminant analysis were employed as statistical classifiers.

Also, for the texture analysis, we used ensemble classification methods, such as random forest (RF) and the error-correcting output codes (ECOC). The random forest method can significantly improve the efficiency of traditional classification algorithms based on decision trees. One of the considered approaches, bagging (RFB), is based on a bootstrap resampling method to randomize the training set and build an ensemble of solutions. Alternatively, the adaptive boosting algorithm (RFAB) is also considered. RFAB is a greedy method for constructing a composition of classifiers, each of which strives to compensate the shortcomings of all previous ones.

The ECOC method is a universal ensemble classifier that allows us to combine a variety of binary classifiers into a multi-class algorithm. In this study, the support vector machine (SVM) with a Gaussian kernel is used as a basic binary learner for ECOC. Thus the method used is abbreviated as ECOC SVM. The effectiveness of ECOC and in particular ECOC SVM was investigated in [9].

Since some of the considered methods are evidently unbalanced, then the number of training samples for each of the considered classes was equalized to provide the equal a priori probability of the classes. We used also the regularized sequential feature selection method proposed in [10] to effectively reduce the feature space.

## 3 Results and discussion

Panchromatic satellite images of very high spatial resolution (0.46 m) obtained with the WorldView-2 equipment were used for the texture analysis. The survey was carried out on the territory of the Savvatyevskoe forestry (Russia, Tver region) at noon on June 25, 2016 at the sun height of 55.7. The survey area, consisting of 2 stitched images with different viewing angles, is indicated by a black frame in Fig. 1.

Numerical calculations to assess the informativeness of statistical textural features were performed for two test regions: Konstantinovsky and Tverskoy Posad - the location of which is shown in Fig. 1. The first of them is located in the eastern part of the Krasnogorsky sandpit near the village of Domnikovo. The region contains several large zones (Fig.2b) corresponding to 5 types of natural and man-made objects with different textures: 1 – buildings, 2 – plantations, 3 – pine forest, 4 – peat bog, 5 – water surface. The size of the moving window was set equal to 91 pixels, since this corresponds to the minimum errors for all considered classifiers. The results of classification using the ECOC SVM method according to the characteristics of the 1st and 2nd orders are shown in Fig. 2c, d, e, f.

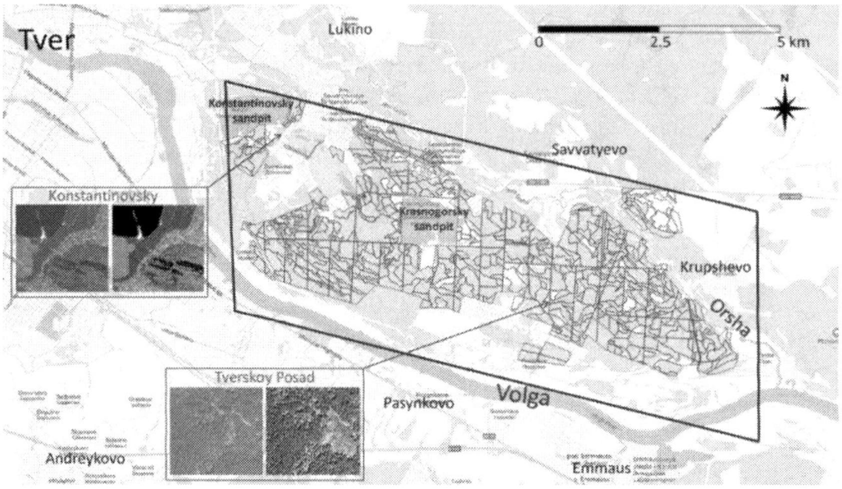


**Fig. 1.** The map of the Savvatyevskoe forestry (Russia, Tver region). Blue contours are boundaries of forest inventory plots, black frame is the WorldView-2 survey area, green arrows show the location of test areas, corresponding enlarged RGB and panchromatic images are shown in green frames.

https://doi.org/10.1051/e3sconf/202022302013

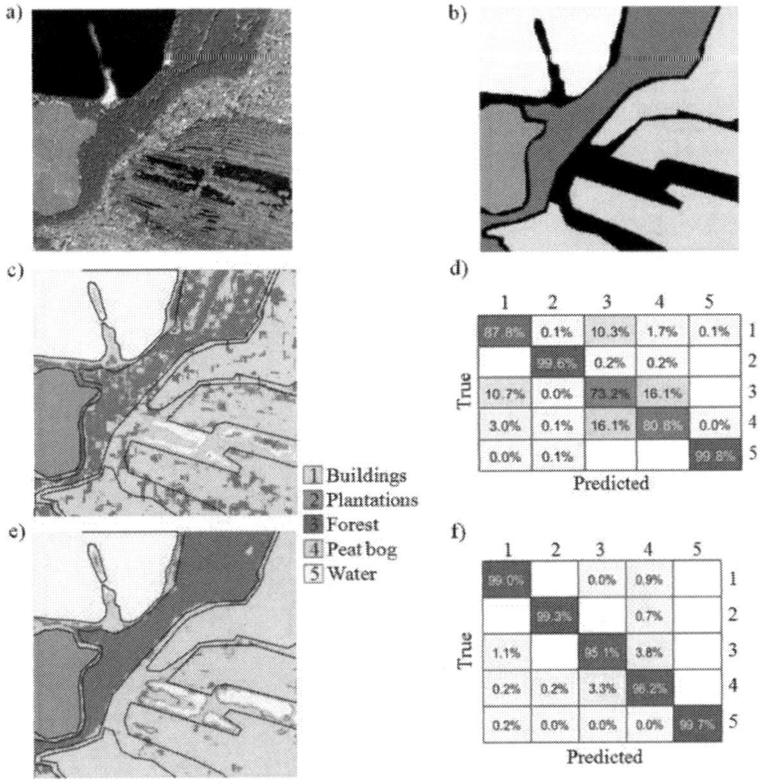


**Fig. 2.** The results of texture classification of the WorldView-2 image of the Konstantinovsky test region using ECOC SVM: a) panchromatic image; b) map of expert classification of objects, c, d) classification by 1st order statistics and the corresponding confusion matrix; e, f) classification by 2nd order statistics and associated confusion matrix.

The test region Tverskoy Posad is located in the mixed forest zone (pine and birch are dominant species) of the Savvatyevskoye forestry and contains stands of various structures: dense canopy, sparse canopy and undergrowth. The contours of the areas corresponding to the listed objects are shown in Fig. 3b. The classification results obtained by using the ECOC SVM method according to the texture characteristics of the 1st and 2nd orders are shown in Fig. 3c, d, e, f. The following terms were introduced as the main characteristics of classification errors: TE is the total probability of error; TOE is the mean omission error and TCE is the mean commission error.

The results of evaluating the classification accuracy obtained for various methods when processing images of the considered test regions are presented in Table 1. The number of training samples (NTS) was chosen equal to 1000 for all calculations. The remaining pixels were used for the independent estimation of classification errors. The number of test samples is much greater than the number of training samples. Estimates of the classification time for one pixel for different NTS values are presented in Table 2.

 https://doi.org/10.1051/e3sconf/202022302013

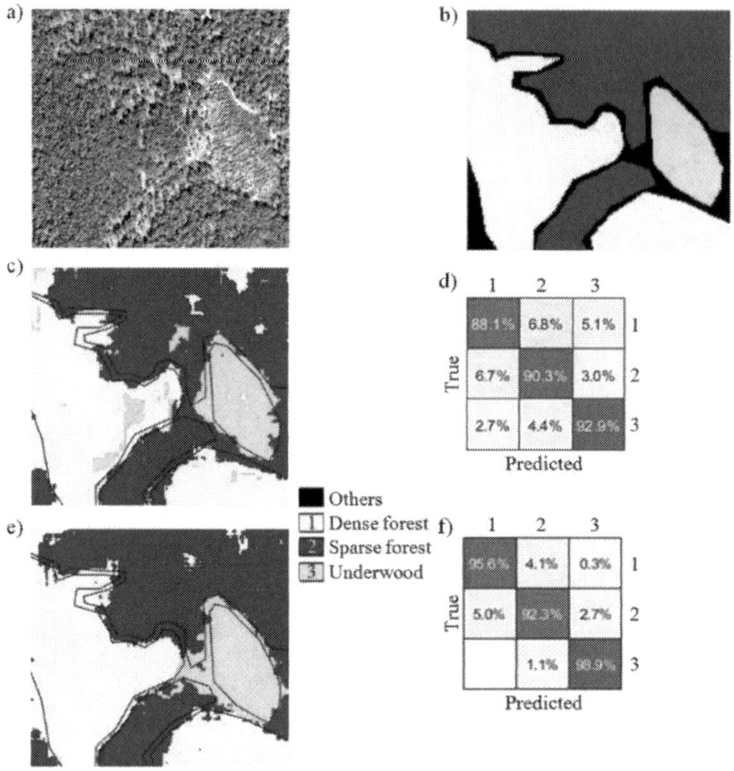


**Fig. 3.** The results of texture classification of the WorldView-2 image of the Tverskoy Posad test region using ECOC SVM: a) panchromatic image; b) map of expert classification of objects, c, d) classification by 1st order statistics and the corresponding confusion matrix; e, f) classification by 2nd order statistics and associated confusion matrix.

The analysis of the results obtained shows that the considered objects can be classified with acceptable accuracy for both test regions. The NC method reveals the lowest accuracy in comparison with all other methods considered in this study, but at the same time it has the highest processing speed, which practically does not depend on NTS and is by 1-2 orders of magnitude less than the speed of ensemble classifiers. The LDA method has almost the same accuracy as NC, but the processing speed decreases by several times. The use of traditional nonlinear classifiers QDA and KNN leads to a significant increase in accuracy, but the processing speed for KNN is less by an order of magnitude.

**Table 1.** Probabilities of classification errors for different methods use texture processing images of the considered test areas. NTS = 1000.

| | | NC | LDA | QDA | KNN | ECOC-SVM | RFB | RFAB |
|---|---|---|---|---|---|---|---|---|
| Konstantynovsky | TE | 0.083 | 0.083 | 0.058 | 0.041 | 0.025 | 0.027 | 0.030 |
| | TOE | 0.075 | 0.073 | 0.049 | 0.035 | 0.022 | 0.024 | 0.026 |
| | TCE | 0.079 | 0.077 | 0.053 | 0.035 | 0.022 | 0.025 | 0.027 |
| Tverskoy Posad | TE | 0.122 | 0.113 | 0.110 | 0.094 | 0.063 | 0.088 | 0.089 |
| | TOE | 0.102 | 0.092 | 0.086 | 0.071 | 0.048 | 0.070 | 0.071 |
| | TCE | 0.174 | 0.155 | 0.159 | 0.143 | 0.097 | 0.140 | 0.141 |

The greatest accuracy can be achieved when using ensemble classifiers. ECOC SVM showed the best results for both test regions. RFB and RFAB methods have a significantly lower processing speed, however, it should be noted that with an increase in NTS, the processing speed for ECOC SVM decreases significantly, as well as for KNN.

**Table 2.** Computational time costs for processing one pixel (microsecond/pixel).

| NTS | NC | LDA | QDA | KNN | ECOC-SVM | RFB | RFAB |
|---|---|---|---|---|---|---|---|
| 100 | 0.15 | 0.55 | 0.57 | 6.56 | 3.52 | 25.06 | - |
| 500 | 0.14 | 0.56 | 0.58 | 13.50 | 5.26 | 25.56 | 24.82 |
| 1000 | 0.10 | 0.59 | 0.57 | 20.44 | 6.65 | 25.24 | 24.74 |

The comparison of the results presented in Fig. 2 and 3 shows that the statistical characteristics of the 1st order are less informative for the classification of the considered types of objects. To reveal more subtle texture differences, it is necessary to use the characteristics of the 2nd order. In this case, it is possible to obtain recognition accuracy of more than 92% for all classes.

The thematic map of natural and man-made objects, recognized on the basis of the automated texture classification of the full image of the Savvatyevskoy forestry is shown in Fig. 4. The ECOC SVM method was used for calculations. Validation of the results obtained based on visual analysis and comparison with ground thematic maps shows good quality of thematic processing based on texture analysis.

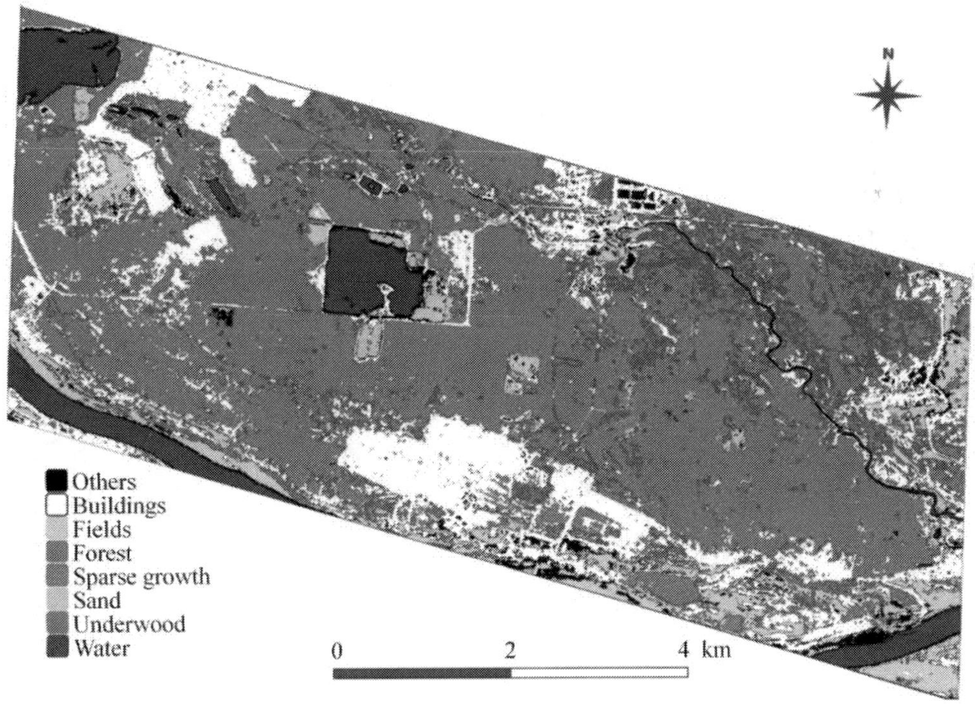


**Fig. 4.** The results of texture classification of natural and man-made objects on the territory of the Savvatyevskoe forestry using WorldView-2 panchromatic images.

## 4 Conclusion.

Texture processing of panchromatic images of WorldView-2 of the territory of Savvatyevskoe forestry (Tver region) was carried out using traditional and ensemble supervised learning classification algorithms. It is shown that the use of 2nd order statistical texture features leads to a significant increase in the classification accuracy and allows solving the recognition problem for a wider class of objects. The use of ensemble algorithms makes it possible to increase the classification accuracy by 2-3 times in comparison with the traditional ones. ECOC SVM reveals the best results and can be considered as the promising effective method, which provides the classification of the natural and man-made objects with an error of less than 3% and the structural features of forest stands with an error of about 6%.

The reported study was funded by RFBR, projects N 20-07-00370 «Fundamental problems of increasing the informativeness of processing data from optoelectronic aerospace devices of high spatial and spectral resolution» and N 19-01-00215 «Investigation of operative opportunities of hyper-spectral technologies of remote sensing of the Earth to solve regional problems using updated hyper-spectral cameras from space».

## References

1. E.V. Dmitriev, A.A. Sokolov, V.V. Kozoderov, H. Delbarre, P.G. Melnik, S.A. Donskoi, Proc. SPIE 11149, Remote Sensing for Agriculture, Ecosystems, and Hydrology XXI, 111491J, (2019)
2. S.E. Franklin, Remote sensing for sustainable forest management (CRC press, 2001).
3. M.P. Ferreira, F.H. Wagner, L.E. Aragão, Y.E. Shimabukuro, C.R. de Souza Filho, ISPRS journal of photogrammetry and remote sensing, **149**, 119-131 (2019)
4. B. Beguet, D. Guyon, S. Boukir, N. Chehata, ISPRS journal of photogrammetry and remote sensing, **96**, 164-178 (2014)
5. J. Chaki, N. Dey, Texture Feature Extraction Techniques for Image Recognition (Springer Singapore, 2020)
6. R.M. Haralick, K. Shanmugam, I. Dinstein, IEEE Transactions on Systems, Man, and Cybernetics, SMC-3, **6**, 610-621 (1973)
7. T. Hastie, R. Tibshirani, J. Friedman, The Elements of Statistical Learning: Data Mining, Inference, and Prediction (Springer Series in Statistics. 2nd ed., 2008)
8. T.G. Dietterich, G. Bakiri, J. Artificial Intelligence Research, **2**, 263-286 (1995)
9. E.V. Dmitriev, V.V. Kozoderov, A.O. Dementyev, A.N. Safonova, Optoelectronics, Instrumentation and Data Processing, **54**, 3, 213–221 (2018)
10. E.V. Dmitriev, Izvestiya, Atmospheric and Oceanic Physics, **50,** 9, 929–942 (2014)

# Fire-induced changes in soil and vegetation in the forest-tundra of Western Siberia

*Oleg* Sizov[1], *Leya* Brodt[2], *Andrey* Soromotin[2], *Nikolay* Prikhodko[2], and *Ramona* Heim[3]

[1]Oil and Gas Research Institute RAS, 119333 3 Gubkina st., Moscow, Russia
[2]Universaty of Tyumen, 625000 6 Volodarskogo st. Tyumen, Russia
[3]Universaty of Munster, 48149 2 Schlossplatz st, Münster, Germany

**Abstract.** Wildfires are one of the main factors for landscape change in tundra ecosystems. In the absence of external mechanical impacts, tundra plant communities are relatively stable, even in the face of climatic changes. In our study, lichen cover was degraded on burnt tundra sites, which increased the permafrost thaw depth from 100 to 190 cm. In old fire scars (burnt 1980 – 1990) of the forest-tundra, vegetation cover was dominated by trees and shrubs. The soil temperature on burnt forest-tundra sites was higher in comparison to conditions of the unburnt control sites and permafrost was was not found at a depth of 2-2,3m. Dynamics of the Normalized Difference Vegetation index (NDVI) from 1986-2020 reveal that immediately after fires, vegetation recovered and biomass increased due to the development of *Betula nana* shrubs. In old fire scars of the forest-tundra (burnt 1980-1990), a significant increase in NDVI values was evident, in contrast to the unburnt tundra vegetation where this trend was less pronounced. We conclude that "greening" in the north of Western Siberia may occur due to fire-induced transformation processes. The role of wildfires in the advance of the treeline to the north, driven by climate change and active economic development of the Arctic, will gradually increase in future.

## Introduction

Wildfires are an important environmental factor, strongly influencing structure and distribution of vegetation, the intensity of the carbon cycle, and properties of the upper soil layer. Recent studies show that fires in the Arctic are one of the main reasons for permafrost degradation [1]. Under natural conditions, fires occur less frequent in tundra than in forest-covered taiga ecosystems [1, 2]. However, the frequency and intensity of tundra fires increase in direct proportion with anthropogenic impact, especially in areas of oil and gas production [3]. Even a single fire leads to strong changes in geochemical and geobotanical conditions of the Arctic landscape, which is characterised by low stability and a high potential of natural resources. Fire-induced changes in water regime, polar vegetation, and tundra and subboreal landscapes of the permafrost zone are the focus of very few studies [1]. We can highlight one study, conducted in the area of the Nadym region of the Yamalo-Nenets Autonomous Okrug [4], which showed that after fire, an intensive morphological change is observed in the upper soil layer, accompanied by a decrease of organic matter, and an enrichment with nitrogen. The impact of fire on forest-tundra in the face of climate warming and greening processes of the Arctic remains poorly studied [5, 6].

To identify, monitor and assess wildfire impacts, remote sensing methods are commonly used, mainly based on remote sensing of the Earth (ERS) of medium and low resolution (Landsat-5,7,8, Sentinel-2,3, TERRA, AQUA, SuomiNPP, etc.) [2]. The handling of large amounts of information allows to estimate the forest-covered area on a global scale and with annual periodicity, using optical as well as radar data [7, 8]. Additionally, a retrospective analysis of changes can be performed, with the possibility to monitor the increase of burnt areas and the advance of shrubs and forest to the tundra (an increase in biomass and an increase in the degree of "greenness") [7]. For retrospective assessments, cloud-based geoplatforms are currently commonly used. Google Earth Engine allows to work with a complete archive of Landsat, MODIS, Sentinel data, as well as to conduct analytical operations and create applied projects (e.g. Global Forest Change, Global Forest Watch) [7]. In Russia, information systems for monitoring forest cover and forest fires based on remote sensing data are also existent [10, 11].

The aim of this work is to identify patterns of soil and vegetation changes in fire scars of different age, situated in the forest-tundra zone in the north of Western Siberia, based on field and remote sensing data.

## Territory and research methods

The territory is efficient within the Nadym region of the Yamal-Nenets Autonomous Okrug (Yamalo-Nenets Autonomous Okrug) and covers the rivers Nyda and Pravaya Khetta (Fig. 1).

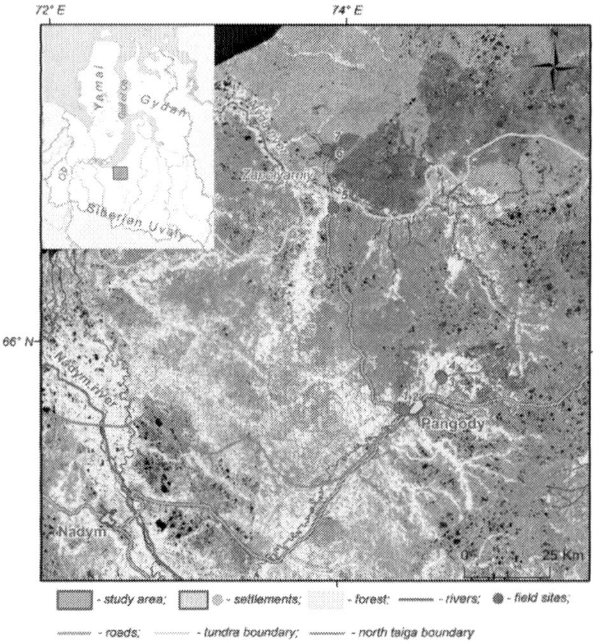


**Fig. 1.** Overview map of the study area (background - Landsat-8, 2018)

The surface relief is even with slight slopes, scattered with many swampy and lacustrine depressions. The upper part of the Quaternary sediments is represented by fluvio-glacial, marine and lacustrine-alluvial sediments. The thickness of organogenic peat deposits is on average 1-1.5 m (up to 5 m in depressions). Height above sea level ranges from 15-900 m. The average annual temperatures are -4.6- (+2)°C, in summer the air warms up to + 14 + 17°C, in winter the average temperature is -24-17°C (weather station Nadym). The study area is located on the border of continuous and discontinuous permafrost. Continuous permafrost occurs in interfluves and within uneven peatlands (temperature -3-1°C). In valleys and on river terraces, the permafrost table is located at a depth of 50-150 m [12].

The study area is situated in the Tazovskoy southern tundra and Nadym-Purskaya forest-tundra provinces of the West Siberian Plain [12, 13].

Plant communities are represented by:

- dwarf birch and willow-dwarf moss-lichen tundra in combination with hummocky fractured-polygonal bogs and lichen-dwarf shrub woodlands (southern tundra formations);
- larch and spruce-larch shrub-moss woodlands in combination with lichen tundra (forest-tundra formations);
- larch and spruce-larch forests with cedar and pine, shrub-moss-lichen forests in combination with dwarf birch-sphagnum-lichen flat and large hillocky bogs (north taiga formations).

The anthropogenic impact is characterised by the development of large oil and gas fields (Medvezhye, Vostochno-Medvezhye). In addition, major gas pipelines (Yamburg-Tula, Yamburg-Yelets, SRTO-Ural, Urengoy-Center, Urengoy-Uzhgorod, etc.) run through the study area.

**Table 1.** Study sites.

| N | Coordinates | Date of field investigations | Characteristic |
|---|---|---|---|
| s.1 | N65°51'03,34" E74°22'54,12" | 16.08.2020 | Unburnt tundra vegetation on the southern border of the forest-tundra zone |
| s.2 | N65°50'51,29" E74°25'01,67" | 16.06.2020 | Forest-tundra site with fire scars of 1990 on the site of earlier burnt areas |
| s.3 | N65°56'13,04" E74°38'45,65" | 17.08.2020 | Forest-tundra site, which burnt before 1985 |
| s.4 | N65°56'05,14" E74°39'16,27" | 18.08.2020 | Forest-tundra site with fire scars until 1985 and fire scar of 2016 |
| s.5 | N66°32'10,5466" E73°53'03,4520" | 17.08.2020 | Forest-tundra site with fire scars before 1988 in the site of earlier burnt areas |
| s.6 | N65°31'19,93" E73°54'17,22" | 17.08.2020 | Southern tundra site with fire scars of 2016 |
| s.7 | N65°32'09,38" E73°52'50,36" | 17.08.2020 | Unburnt tundra vegetation at the southern border of the tundra zone |

To identify the characteristics of soil and vegetation recovery in the fire scars of different age in the forest-tundra zone, investigations were carried out at seven sites, which include both unburnt control sites and fire scars of different age (Fig. 1, Table 1). In each of the studied sites, we conducted geobotanical vegetation mapping and measured soil temperatures at different depths. Soil temperature was measured with a Tr 46908 thermometer (TR di Turoni & c. Snc, Italy) and drilling was carried out using a hand-held motor-drill Stihl BT 360 (Stihl, Germany).

To assess the rate of vegetation recovery for each site, we calculated the Normalized Difference Vegetation Index (NDVI) from 1986-2020, using the Landsat data harmonization feature of Google Earth Engine. As study period, we chose the interval from June 30 to August 31. All available images in that period, with a quality threshold of more than 50%, were used to calculate the median NDVI value.

## Results and discussion

We found significant differences in vegetation and soil temperature (Fig. 2) between the studied sites.

Site s.1 is characterised by unburnt tundra vegetation and is situated at the southernmost border of the forest-tundra zone in our study area. *Betula nana* dominates in the shrub layer (35% of the total cover), *Vaccinium uliginosum* is the dominating the dwarf shrub species (15% of the total cover), and a large part of the site is covered with *Cladonia stellaris* (more than 75% of the total cover). On this site, soil temperature was found to be negative (-0.3°C) at a depth of 210 cm. On the second unburnt control site, situated at the southern border of the tundra zone (s.7) 85% of the total cover was occupied by *Cladonia stellaris*. Permafrost was found at a depth of 120 cm (soil temperature of -0.5°C) and an ice lens (5 cm thick) was revealed by the borehole.

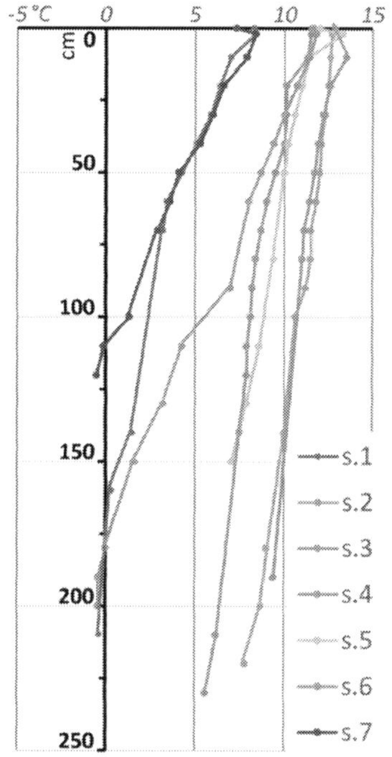


**Fig. 2.** Soil temperature at different depths of all seven site

In the site of the recently burnt fire scar (s. 6), burnt in 2016, *Polytrichum commune* occupies 70% of the total cover, and there are single specimens of *Betula nana, Salix arctica*, and *Ledum palustre*. The soil temperature in the upper part of the borehole is 3-4°C higher compared to the unburnt control conditions. Permafrost was found only at depths of 185-195 cm.

In the forest-tundra site with the fire scar of 1990 (s.2), *Betula nana* occupies the main part of vegetation cover (40% of the total cover), there is a single specimen of *Betula tortuosa*, and the moss-lichen cover is practically absent. The minimum soil temperature at a depth of 210 cm was 8.1°C. On the site of the burnt forest-tundra, where the fire event took place before 1985, *Vaccinium uliginosum* dominates the vegetation (s.2) (70% of the total cover) and the moss-lichen layer occupies 2% of the total cover. Soil temperature at a depth of 210 cm amounts to 6.2°C. Vegetation on the site in the northernmost part of the forest-tundra, which burnt before 1988 (p. 5), is characterised by *Betula nana* (40% of the total cover) and *Cladonia stellaris* (18% of the total cover). At a depth of 153 cm, the soil temperature was 7.1 ° C. *Vaccinium uliginosum* (40% of the total cover) and *Cladonia stellaris* (53% of the total cover) dominate the vegetation cover on the site of multiple fire events (before 1985 and in 2016). At a depth of 190 cm, the soil temperature was 9.4°C.

Median NDVI values for 1986-2020 show that in the unburnt control sites, the biomass is significantly lower than in the burnt sites (Fig. 3). In addition, in sites s.4 and s.6, the downward slopes, related the fires of 2016, are clearly visible. It can be noted that a gradual increase in NDVI values is evident in old fire scars at least since the late 1990s.

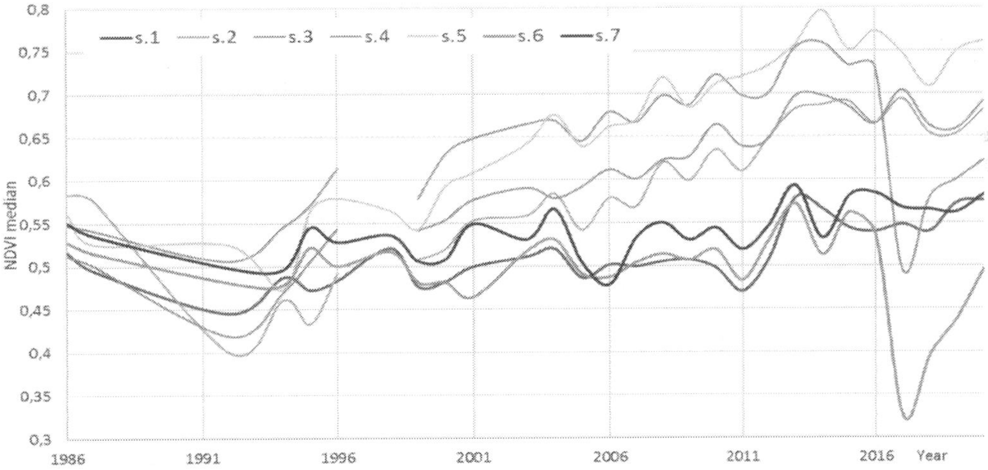


**Fig. 3.** Median NDVI values of tundra vegetation for fire scars of different age and unburnt control sites between 1986-2020 (based on Landsat data in Google Earth Engine).

In general, the unburnt control sites (s.1 and s.7) are characterized by the dominance of lichens (more than 70% of the total cover). At the same time, *Betula nana* occupies a significant proportion of shrub cover. On the recently burnt fire scar of 2016 (s.4), litter is characterised by burnt remnants, and lichen cover is practically absent (less than 5% of the total cover). At the same time, shrubs and dwarf shrubs show enhanced growth. On the burnt site within the tundra zone (s.6), *Polytrichum commune* occupies a large share of the site.

Permafrost conditions show clear differences. In the unburnt control sites s.1 and s.7, there is a sharp decrease in soil temperature to 2–3°C immediately below the surface, already at a depth of 100 cm. For the tundra site with a recent fire scar (s.6), an increase in the soil temperature of the upper part of the soil strata was found in comparison with the unburnt control site by 4-5°C, while the depth of the permafrost is comparable to the unburnt control site (s.1). For the site of multiple burns, soil temperature is significantly higher, and the gradient of its decrease is less compared to the unburnt control site. Permafrost is only detected at depths below 2-2.3 m.

NDVI dynamics from 1986-2020 show two main features:

- vegetation recovery and an increase in NDVI values are observed after fire, mainly due to shrub encroachment (*Betula nana*) (areas s.4 and s.6);
- for forest-tundra sites with old fire scars (1980-1990), NDVI values increase significantly in comparison to the unburnt control tundra vegetation, where this trend is

much less pronounced. I.e. "greening" of the tundra on a local scale occurs, in particular, due to fire-induced changes.

In many studies, conducted on the territory of Yakutia and Alaska, a recovery of the permafrost and the original tundra communities, on fire scars, is noted. However, there are also examples for biomass growth, shrub encroachment and the development of forests in tundra ecosystems after fires [14, 15, 16]. In this study, we found a clear tendency of tundra vegetation towards shrub encroachment and the development of forest in the forest-tundra zone as a result of fire-induced effects. At the same time, the degradation of permafrost and an increase in soil temperature does not occur immediately after a fire, but gradually, as the root system of shrubs and trees develops. In this case, the fire acts only as a primary factor in disrupting the protecting effect of the lichen cover.

### Conclusion

Tundra plant communities in the north of Western Siberia are characterised, in the absence of external mechanical influences, by high stability, despite the observed climatic changes. The main factor for the possible replacement of tundra plant communities by secondary plant communities are natural fires. Post-fire succession leads to further degradation of the permafrost due to the development of the root system and an increase in the thermal conductivity of soils under conditions of a steady increase in average annual air temperatures. Our results confirm the presence of shrub expansion and a decrease of lichen cover after fires, both in short- and long-term recovery after fire of the forest-tundra in the north of Western Siberia. Within the face of climate change and the economic development of the Arctic, the role of wildfires in the tundra as one of the factors of landscape change will increase. Recent developments in remote sensing make it possible to develop an objective system for long-term fire monitoring and the assessment of post-fire changes in vegetation, in order to prevent negative ecological consequences.

This study was funded by RFBR according to the research project № 20-55-71004 and Belmont Forum project № 1729 SERUS.

## References

1. I. Nitze, G. Grosse, B.M. Jones, V.E. Romanovsky, J. Boike. Remote sensing quantifies widespread abundance of permafrost region disturbances across the Arctic and Subarctic, *Nature Communications,* **Vol. 9 No. 1**, pp. 1–11 (2018)
2. A.V. Rocha, M.M. Loranty, P.E. Higuera, M.C. Mack, F.S. Hu, B.M. Jones, A.L. Breen, E.B. Rastetter. S.J. Goetz, G.R.Shaver. The footprint of Alaskan tundra fires during the past half-century: implications for surface properties and radiative forcing, *Environmental Research Letters*, **Vol. 7, No. 4**, pp. 1–8 (2012)
3. M.K. Raynolds, D.A. Walker, K.J. Ambrosius, J. Brown, K.R. Everett, M.Kanevskiy, G.P. Kofinas, V.E. Romanovsky, Y. Shur, P.J. Webber/ Cumulative geoecological effects of 62 years of infrastructure and climate change in ice-rich permafrost landscapes, Prudhoe Bay Oilfield, Alaska, *Global Change Biology*. **Vol. 20, No. 4**, pp. 1211–1224 (2014)
4. E. Abakumov, A. Pechik, E. Chebyakina(Maximova), G. Shamilishvili. Effect of the Wildfires on Sandy Podzol Soils of Nadym Region, Yamalo-Nenets Autonomous District, Russia. *Applied and Environmental Soil Science*, **Vol. 2020** (https://doi.org/10.1155/2020/8846005)

5. P.D. Jones, D.H. Lister, T.J. Osborn, C. Harpham, M. Salmon, C.P. Morice. Hemispheric and large-scale land-surface air temperature variations: An extensive revision and an update to 2010, *Journal of Geophysical Research: Atmospheres*, **Vol. 117, No. D5**, pp. 1–29 (2012)
6. J.A. Sobrino, Y.Julien, S.García-Monteiro. Surface Temperature of the Planet Earth from Satellite Data, *Remote Sensing,* **Vol. 12, No. 2**, pp. 1–10 (2020)
7. M.C. Hansen, P.V. Potapov, R. Moore, M. Hancher, S.A. Turubanova, A. Tyukavina, D.Thau, S.V. Stehman, S.J. Goetz, T.R. Loveland, A. Kommareddy, A. Egorov, L.Chini, C.O. Justice, J.R.G. Townshend. High-Resolution Global Maps of 21st-Century Forest Cover Change, *Science*, **Vol. 342, No. 6160**, pp. 850–853 (2013)
8. M. Shimada, T. Itoh, T. Motooka, M. Watanabe, S. Tomohiro, R. Thapa, R. Lucas. New global forest/non-forest maps from ALOS PALSAR data (2007–2010), *Remote Sensing of Environment*, **Vol. 155**, pp. 13–31 (2014)
9. H. Epstein, U. Bhatt, M. Raynolds, D. Walker, J. Pinzon, C.J. Tucker, B.C. Forbes, T. Horstkotte, M. Macias-Fauria, A. Martin, G. Phoenix, J. Bjerke, H. Tømmervik, P. Fauchald, H. Vickers, R. Myneni, T. Park, C. Dickerson. Tundra greenness, *Bulletin of the American Meteorological Society*, 2018, **Vol. 99, No. 8**, pp. 165–169.
10. E.A. Loupian, S.A. Bartalev, I.V. Balashov, V.A. Egorov, D.V. Ershov, D.A. Kobets, K.S. Sen'ko, F.V. Stytsenko, I.G. Sychugov. Sputnikovyi monitoring lesnykh pozharov v 21 veke na territorii Rossiiskoi Federatsii (tsifry i fakty po dannym detektirovaniya aktivnogo goreniya) (Satellite monitoring of forest fires in the 21st century in the territory of the Russian Federation (facts and figures based on active fires detection)), *Sovremennye problemy distantsionnogo zondirovaniya Zemli iz kosmosa (Current problems in remote sensing of the Earth space)*, **Vol. 14, No. 6**, pp. 158–175 (2017)
11. I.V. Balashov, A.V. Kashnitskii, S.A. Bartalev, S.S. Bartalev, M.A. Burtsev, I. I. Vorushilov, V.A. Egorov, V.O. Zharko, D.A. Kobets, A.M. Konstantinova, E.A. Loupian, I.A. Saigin, K.S. Senko, F.V. Stytsenko, I.G. Sychugov, S.A. Khvostikov, T.S. Khovratovich. VEGA-Les: information system for complex monitoring of forests and hunting grounds in Russia. *Sovremennye problemy distantsionnogo zondirovaniya Zemli iz kosmosa (Current problems in remote sensing of the Earth space)*. **Vol. 17, No. 4**, pp. 73-88 (2020)
12. Atlas Yamalo-Nenetskogo avtonomnogo okruga (Atlas of the Yamal-Nenets Autonomous Okrug), Moscow: GUGK, 248 p (2004)
13. I.S. Il'ina, E.I. Lapshina, N.N. Lavrenko Rastitel'nyi pokrov Zapadno-Sibirskoi ravniny (Vegetation of the West Siberian Plain), *Novosibirsk: Nauka*, 248 p (1985)
14. R.E. Hewitt, T.N. Hollingsworth, F.S. Chapin, D.L. Taylor. Fire-severity effects on plant-fungal interactions after a novel tundra wildfire disturbance: implications for arctic shrub and tree migration, *BMC ecology*, **Vol. 16**, pp. 1–25 (2016).
15. S.M. Landhausser, R.W. Wein, Postfire Vegetation Recovery and Tree Establishment at the Arctic Treeline: Climate-Change-Vegetation-Response Hypotheses, *Journal of Ecology*, **Vol. 81, No. 4**, pp. 665–672 (1993)
16. R. J.Heim, A.Bucharova, L.Brodt, J.Kamp, D.Rieker, A.V. Soromotin, ... & N.Hölzel. Post-fire vegetation succession in the Siberian subarctic tundra over 45 years. *Science of The Total Environment*, 143425 (2020)

# Analysis of floodable areas of the Angara river

*Gachenko* Andrey[1,*] , *Hmelnov* Alexey[2]

[1]Irkutsk Scientific Center of the Siberian Branch of the Russian Academy of Sciences (ISC SB RAS), Irkutsk, Russia
[2]Matrosov Institute for System Dynamics and Control Theory of Siberian Branch of Russian Academy of Sciences (IDSTU SB RAS), Irkutsk, Russia

**Abstract**. In this work, the authors present a technology for riverside terrain model building that has been tested on a number of scientific projects to study the littoral area of tail race of the Irkutsk Hydroelectric Power Station and the Bratsk Reservoir. This model is used for forecasting changes in the reservoir shorelines associated with wastewater in the cascade of hydroelectric power stations. The technology described in the work was approved to solve a number of practical problems and showed its effectiveness. Specialized application software was developed and terrain data from various sources were used to specify and detail the end result.

The Angara is a river in Eastern Siberia, the largest right feeder of the Yenisei, the only river flowing out of Lake Baikal. The Angara flows on the territory of the Irkutsk region and Krasnoyarsk territory.

The length of the Angara river is 1779 km, the catchment area of the river – 1039000 square meters. Total river fall is 380 meters, its slope is 0,2 m/km.

Normal annual water flow at the source of the Angara is 1950 m3 / sec. Water consumption in the Angara remains high throughout the year and that is very important for the power system. As a result of the flow control by the Irkutsk, Bratsk and Ust-Ilimsk water storage basins, a redistribution of water consumption took place: in the period of shallow water they significantly increased, and during high water periods they decreased.

With a relatively short length, the Angara has a significant drop equal to 380 m and a large hydropower potential.

On the banks of the Angara there are such large cities as: Irkutsk, Angarsk, Usolye-Sibirskoye, Svirsk, Bratsk, Ust-Ilimsk, Kodinsk and many villages and towns.

The most dangerous area in terms of potential flooding is the zone from the dam of the Irkutsk hydroelectric power station (IHES) to the city of Angarsk with a length of 67 km. About 1 million people live here and the largest enterprises of the Irkutsk region are located. Taking into consideration high risks of urban areas flooding with increased discharges through the stream gauges of Irkutsk power station, it is required to build models to determine the potential zones and boundaries of flooding. Regular space monitoring of

* Corresponding author: gachenko@icc.ru

the required area makes it possible to clarify the boundaries of coastlines with an accuracy of several meters, but it is quite costly affair.

With a sufficiently precise relief model it is possible to estimate potential flooding zones and boundaries, taking into account discharges at the Irkutsk HES. In addition, taking into account the period of extreme low water in the basins of Lake Baikal and the river Angara [3] that started in 2014, there was a problem of shallowing boundaries obtaining of the tail-bay of the IHES and the riverside areas of the Bratsk reservoir.

As a result, it became necessary to develop a relief model that takes into consideration the shape of the river bottom.

Using the modern GIS-technologies: QGIS and original software (TINSmith, Fig. 1), with the Delaunay triangulation [1], we modelled zones of the possible flooding during the rise of water level in the HPS water cascade of the Angara River. We identified dangerous sites and settlements threatened under various scenarios of the water content [2, 3].

For these methods, the topography from the open sources, SRTM and ALOS, pilot maps and vector map on a scale of 1:50000 were also used. Due to significant inaccuracies of measurements from the open sources near the coastline of water bodies, we decided not to use them and take the terrain from vector topographic maps as a basis.

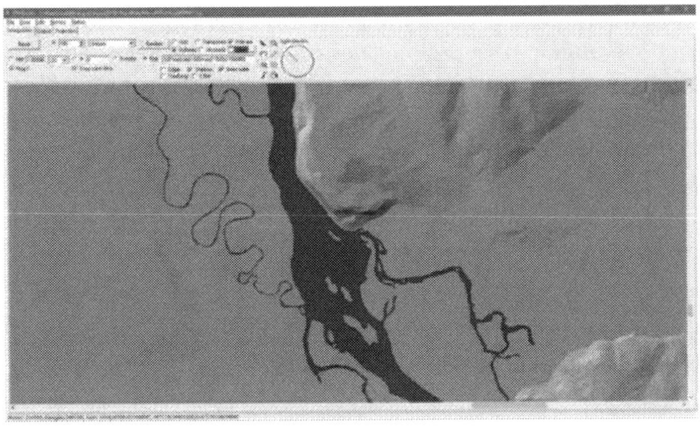

**Fig. 1.** Special TINSmith software for the Delaunay triangulation.

To generate a topographic model, various layers of the topographic base are used, which can play a different role in triangulation that describes topography. All this is adjusted in a special way in the TINSmith original software: horizontal and vertical control (topography), surfaces of reservoirs (surfaces with a constant height), rivers, and streams (linear and areal objects) are used to correct the topography., the marks of water edges and data on the underwater topography are used in the work with the underwater part. We implemented an algorithm for correcting the topography generated from contour lines. This enables us to correct errors typical of these algorithms: artefacts that create a terrace effect.

At first, we take an underwater topography (the triangulation correction algorithm). We have proposed a technique for terrain and underwater topography, which suggests a separate construction of the underwater topography triangulation. For this, an algorithm of horizontal sites is used. After that, the points and rigid edges of the resulting triangulation are saved to files. These files are used in a combined triangulation. Therefore, the edges added to the triangulation for correction will be included in the final combined triangulation and will correct the stair stepping (terrace) effect. We cannot directly apply the triangulation corrections to mixed triangulation because after subtracting the depth from the water level, horizontal triangles become non-horizontal.

For the coordinated combining the contours of the terrain coastline with the vectorized coastline of underwater topography, the problem of matching these lines (topographies) was solved because the underwater topography was obtained from one data and the terrain from other. To solve the problem of verifying reliefs, the model of obtaining a consistent topography was implemented based on the following stages:

1. obtaining source data (vectorization of the printed version of the map in the absence of an electronic version);
2. conversion of the vector topographic base and digitized pilot map to a single coordinate system (WGS 84);
3. setting morphing parameters and conducting morphing itself;
4. obtaining a terrain coordinated with underwater topography.

During morphing (a mechanism for transforming and combining underwater topography with terrain), there is a continuous plane transformation that can combine inaccurate contours of the coastlines with more accurate ones. To combine maps with various types of topography, layers of the coastlines located both in underwater topography and terrain are used. The transformation is applied to various layers of the underwater topography map (isobaths and depth marks). The data resulting from morphing appear to be consistent with the information about terrain.

The contour combining (the problem arises from the inconsistency of terrain with underwater topography) was implemented as a special task in the IrkGV electronic map viewer. The operator working with the program adjusts the parameters in a certain way, which combine the points of the source (inaccurate) and target (accurate) maps. When the contours are combined, the start and end points of the arrows drawn by the user are projected onto the contour points of the corresponding maps nearest to them. Next, a mapping of each contour section of the object on the source map is generated into a specified object on the target map for each pair of arrows that are adjacent on both contours.

The points of contour sections that are combined with each other are displayed through a linear transformation of the parametric coordinates of the curves. If the result obtained is not suitable, then to guarantee the combining some characteristic points of contours, it is necessary to compare these points with each other. After entering information about the combining maps, a file of displacements is formed, in which, in addition to the obviously specified displacements, there are displacements calculated for intermediate points of the contours.

The generation of the topography of the Angara River fragment for a section of 135 km in the interval from Irkutsk HPS to the town of Svirsk and the Irkutsk Reservoir resulted in obtaining a comprehensive digital model (Fig. 2) of the underwater topography combined with the terrain. A digital elevation model of this kind has been obtained for the first time and has no analogues.

**Fig. 2.** Triangulation for a given area and an example of possible flooding under various scenarios of the water content (the Irkutsk Reservoir). Contour lines are plotted every 20 cm.

Based on the modelling various scenarios of water level rise using the QGIS standard methods, we obtained an album with maps of the settlements (103 objects) that are subject to flooding. A map with contour lines was generated for the level rise of up to 5 m and contour lines every 20 cm (Figs 2-4). An expert assessment of the risks of the possible flooding was carried out for each settlement.

Based on the technique described above, we developed digital elevation models of the Irkutsk, Bratsk and Ust-Ilimsk reservoirs (Figs 2-4), including the zones of possible flooding (in the form of contour lines) under various scenarios of the water content. Based on this technique, we generated an electronic map of the coastline (buffer zone from 4 to 10 km) and obtained a digital album for the Angara River within the Irkutsk Region. We processed 103 settlements within the Irkutsk Region and constructed contour lines with flooding threat levels for each of them.

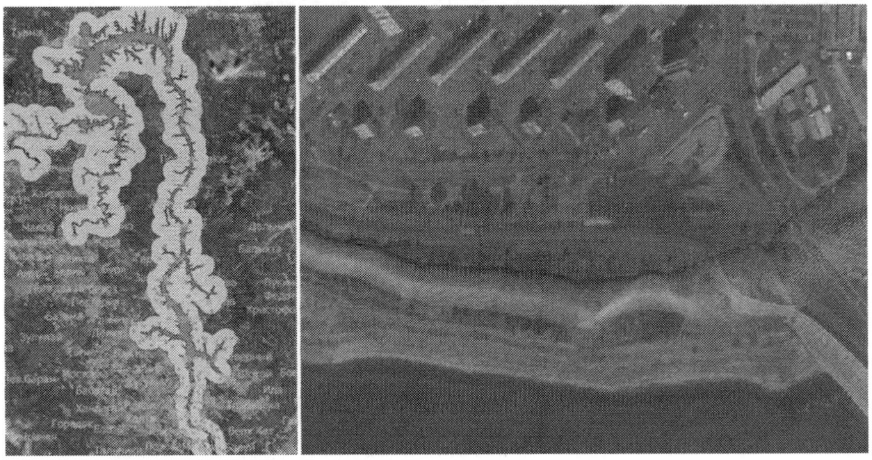

**Fig. 3.** The Bratsk Reservoir and a fragment of the Bratsk city.

**Fig. 4.** The Ust-Ilimsk Reservoir and the fragment of the Ust-Ilimsk town.

In conditions of elevated and extremely high influx to Lake Baikal, taking into account the current rules for regulating water levels in the lake, there are high risks of flooding in the downstream of Irkutsk HPS. To assess the possible damage, we have developed a technique for modelling the channel of the Angara River as well as special software components that allow us to investigate flooding zones at various discharges through HPS.

Using this technique and the developed digital elevation model, we can make various forecasts about the scenarios and threats under different water level regimes for all settlements adjacent to the bank of the Angara River in the section from the river source to the Bratsk Reservoir. We have identified dangerous sites and settlements that are threatened under various scenarios of the water content.

*The work was carried out with the support of RAS (projects:AAAA-A17-117032210079-1, AAAA-A19-119111990037-0) and framework of the block of the ISC SB RAS integration program 4.1.2.*

## Reference

1. A. V. Skvortsov 2002 *Delaunay triangulation and its appliance* (Tomsk: Tomsk University press) p 128.
2. I. V. Bychkov and V. M. Nikitin *Control of level of Lake Baikal: problems and possible issues* Geography and environmental resources **3** pp 5–16 (2015).
3. N. V. Abasov, M. V. Bolgov, V. M. Nikitin and E. N. Osipchuk *On regulation of the urovenny mode of Lake Baikal* Water Resources **44** pp 407-416 (2017).

# USE OF SPECTRAL SURFACE CHARACTERISTICS FOR MAPPING SOIL COVER STRUCTURE UNDER KRASNOYARSK FOREST-STEPPE CONDITIONS

*Valentina* Latysheva[1], *Tatyana* Demyanenko*[1], *Irina* Botvich[2], *Dmitriy* Emelyanov[2], and *Sergey* Khizhnyak[1]

[1] Federal State Budget Educational Institution of Higher Education Krasnoyarsk State Agrarian University, 90 Mira Avenue, Krasnoyarsk City, 660049, Russia
[2] Institute of Biophysics SB RAS, Krasnoyarsk 660036, Russia

**Abstract.** The relations between the spectral surface characteristics of the elements of the soil cover structure and soil properties in the Krasnoyarsk forest-steppe of Central Siberia were investigated. It was revealed that the most informative parameters for field spectrometry are the content of humus, carbonate carbon dioxide and the prevailing particle-size fractions. A statistically significant relationship between the elements of the soil cover structure and the reflectivity of soils has been confirmed by means of multidimensional statistics. The wave lengths with the greatest coupling force are highlighted. Regression equations for remote study of soil cover structure have been obtained, which can be used if additional point studies are carried out in a wider range of test parameters. *Keywords: soil cover structure, spectral brightness coefficient, humus, particle-size distribution, multiple regression.*

## Introduction

Sustainable and environmentally sound crop production requires consideration of all factors, including the heterogeneity of the soil cover or its structure (SCS ) [1]. The most optimal method of accounting is digital mapping of the SCS with the subsequent use of a digital map in the adaptive landscape agriculture system [2]. Large-scale soil mapping is extremely economical and labor-intensive due to the specifics of the mapped object. The use of remote sensing data (RSD) greatly facilitates work, but requires a detailed methodological framework in a specific region.

Agricultural land on the territory of the Krasnoyarsk Region occupies about 5 million hectares, but at the moment the current soil and cartographic support of this area is extremely limited, point-wise. The development of methodological possibilities of using RSD for the recognition of soil heterogeneities is at the initial stage of development and is presented by single studies [3, 4]. The purpose of this work is to identify the relations

* Corresponding author: t-demyanen@mail.ru

between the spectral characteristics of the soil surface and its physicochemical parameters in the conditions of a pronounced microrelief of arable land of the Krasnoyarsk forest-steppe.

## Objects and Methods

The study was carried out on a test site with an area of 10 hectares in the Minderlinskoye experimental-industrial farm of the Sukhobuzimsky district of the Krasnoyarsk Region (56˚25, 92˚53), on which a map of the SCS was previously drawn up.

A semi-closed intermountain basin, which is the Krasnoyarsk forest-steppe, stretches in a narrow strip along the left bank of the Yenisei with a hilly- ridged plain, the morphological structure of which is complicated to its periphery. From the east, the basin is bounded by the Yenisei Ridge, from the south by the spurs of the Eastern Sayan, from the southwest by the Kemchuga Highlands, and in the north and northwest it has no clear borders and merges with the taiga. In tectonic terms, this is a relatively stable section of the Kemchug depression [5, 6]. The territory is composed of sandstones and siltstones of the Jurassic system, covered by loess-shaped and deluvial loams and clays. The wavy denudation relief of the territory is represented by gentle (up to 15 °) slopes, the upper parts of which are formed under the influence of plane flushing, the middle ones - by the joint action of gravitational and fluvial processes. Long-term seasonal permafrost causes the formation of a pronounced microrelief - suffosion-subsidence on watersheds and extended on slopes.

The experimental field is located on the eastern slope of the watershed section between the Buzim-Minderla River. The absolute height difference is 10 meters (Fig. 2). The surface of the site is characterized by a pronounced hummocky-depression microrelief with open and semi-closed shallow stretches.

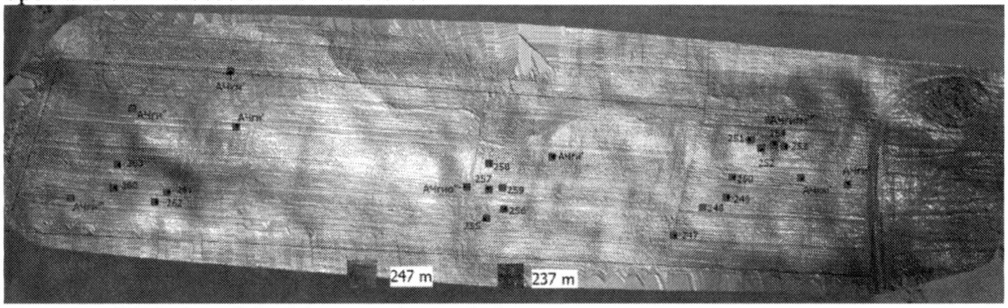


**Fig. 1.** Digital model of the experimental site

The climate of the Krasnoyarsk forest-steppe is characterized as very continental (continental coefficient calculated according to Ivanov - 220) and semi-humid . The relatively sharp continentality of the region is attributed to the great distance from the seas and the ocean and, as a result, a large temperature difference between the coldest (up to -49.4 ˚C) and the warmest (up to 36.1 ˚C) months of the year. Winters are long and cold, summers are short and hot, often dry. The sum of the active temperatures is 1886 ° C, the average annual air temperature is 0.4 ˚C. In spring, the temperature increases rapidly, in autumn it also drops rapidly. There is a significant risk of late spring frosts returning. The frost-free period ranges from 85-120 days. The duration of the active vegetation period is 108 days [7].

The average annual rainfall is 400 mm, most of which falls on the period of active vegetation. The average height of the snow cover is 21.5 cm. The hydrothermal coefficient is 1.2. The humidification coefficient (0.81) characterizes the test zone as semi-humid with a periodically flushing type of water regime. The soils of the Krasnoyarsk forest-steppe belong to the cold (long-term-seasonally freezing) type.

The soil cover of the study site is represented by spotting of typical clay-illuvial agrochernozems, podzolic and hydro-metamorphised with a humus horizon thickness from low to high and low-power cryogenic-micellar agrochernozems.

In May 2019 (during this period, the soil surface was open), 153 point samples were taken from the arable soil horizon of the site with a parallel spectral survey of the soil surface by a field spectrometric complex consisting of a Spectral Evolution PSR-1100F spectrometer with the possibility of internal data storage and a calibration reflective element. Remote data collection and analysis was carried out using the DARWin SP Data Acquisition Package joint software. The height of the survey is100 cm above the soil surface, the area is 50 cm2. The shooting was carried out in clear weather from 11:00 to 15:00. The position of the spectrometer was installed vertically above the surface of the shooting area (in nadir). Particle-size analysis was carried out in the selected samples, the content of humus and carbonate carbon dioxide [8] was determined. Mathematical processing of the results was carried out with the combined use of the programs Microsoft Excel and StatSoft Statistica 8.0.

## Results and discussion

The humus content in the analysis of point-like samples ranged from 11 to 5%, which is estimated as very high, high and medium. The humus horizon, despite its small power, is very fertile and highly humus. In some areas, a reduced content of humus is observed, due to the manifestation of erosion processes. The calcium carbonate content is small and does not exceed 0.8%.

The particle-size distribution of point samples is heavy: from heavy loamy to medium loamy,, measured in every third sample. The predominant fractions are coarse dust and silt. There is a significant depletion of the sand fraction.

Spectral brightness coefficients were obtained in the range from 320 to 1100 nm. For convenience of work, the available spectral range by means of factor analysis of multidimensional statistics was reduced to one generalized index (factor) in 11 categories on the basis of division of spectrum of electromagnetic radiation into groups [9]. With the help of discriminant analysis, only those in which the soil parameters differ significantly in the spectral brightness set were identified from the selected areas (Table 1). To group the data, the variable "Шифр - Code" is selected, which means the grouping of points by contours in the structure of the soil cover.

**Table 1.** Discriminant Analysis Results.

| Range wave lengths | Lambda Wilks | Private lambda | Value F-criterion | P-levels |
|---|---|---|---|---|
| 320-380 | 0,511696 | 0,793567 | 7,387798 | 0,0000035 |
| 451-480 | 0,564171 | 0,719755 | 11,0579 | 0,0000052 |
| 481-510 | 0,51822 | 0,783576 | 7,844115 | 0,0000015 |
| 621-700 | 0,482611 | 0,841391 | 5,353615 | 0,000154 |
| 701-770 | 0,472976 | 0,858531 | 4,679756 | 0,000551 |
| 771-820 | 0,456788 | 0,888957 | 3,547559 | 0,004716 |

Using the same factor analysis within each range, we identified specific wave lengths according to the value of the correlation coefficient with the highest binding strength: 360,

465, 495, 660, 733 and 800 nm, and then to establish the relationship between soil parameters and spectral properties of soils used the coefficients of spectral brightness in the given lengths.

The dependence of reflectivity on the studied soil parameters was established by multiple regression analysis (Table 2).

**Table 2.** Multiple Regression Analysis Results

| Soil parameter, % | Multiple R | P-levels | Wave length, nanometer | Regression coefficient |
|---|---|---|---|---|
| Sludge fraction | 0,5737 | 0,000279 | 360 | 35,733 |
| | | | 46.5 | -160,425 |
| | | | 495 | 123,651 |
| Coarse dust fraction | 0,5373 | 0,001043 | 360 | -38,0765 |
| | | | 465 | 133,4074 |
| | | | 495 | -99,1852 |
| Physical clay | 0,6333 | 0,000021 | 360 | 46,041 |
| | | | 465 | -169,569 |
| | | | 495 | 126,365 |
| Humus | 0,5294 | 0,000001 | 360 | 10,9312 |
| | | | 465 | -33,8066 |
| | | | 495 | 24,5224 |
| $CaCO_3$ | 0,5891 | 0,000001 | 465 | 2,20776 |
| | | | 495 | -2,67923 |
| | | | 660 | 0,94665 |
| | | | 800 | 0,35709 |

A statistically significant relationship of particle size distribution exists only in the predominant fractions: coarse dust, slit and physical clay with wave lengths of 360, 465 and 495 nm in the ultraviolet, blue and blue parts of the spectrum, respectively. The relationship between humus and spectral brightness coefficients is observed in the same ranges (360, 465 and 495 nm). Carbonates of calcium show bonds with the light blue, blue, red and near infrared areas of a range in lengths of 465, 495, 660 and 800 nanometers, respectively.

The regression coefficients are statistically significant for all soil parameters, but their regression equations cannot be applied in practice due to the large error of the results (Figure). In the case of calcium carbonates and humus, such an error is minimal, with a particle size distribution of maximum.

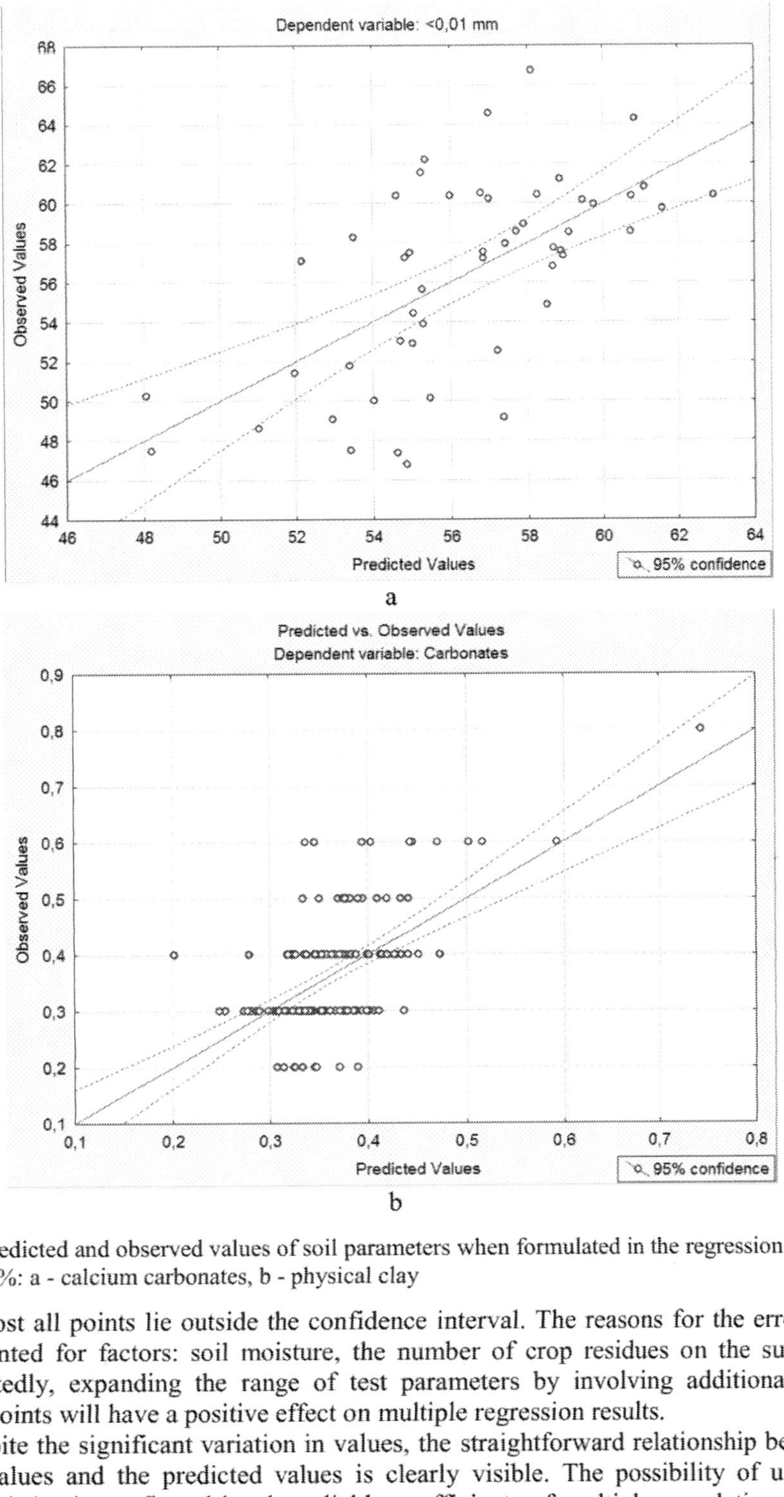


**Fig. 2.** Predicted and observed values of soil parameters when formulated in the regression equation,%: a - calcium carbonates, b - physical clay

Almost all points lie outside the confidence interval. The reasons for the error may be unaccounted for factors: soil moisture, the number of crop residues on the surface, etc. Undoubtedly, expanding the range of test parameters by involving additional off-field testing points will have a positive effect on multiple regression results.

Despite the significant variation in values, the straightforward relationship between the actual values and the predicted values is clearly visible. The possibility of using these characteristics is confirmed by the reliable coefficients of multiple correlation (Table 2), however, in order to improve the accuracy of the forecast, it is necessary to develop the data analysis methods.

## Conclusion

Thus, the most informative of the studied parameters for field spectrometry are the content of humus, calcium carbonates, physical clay and the predominant particle-size fractions. A statistically significant relationship between the elements of the soil cover structure and the reflectivity of soils has been confirmed by means of multidimensional statistics. Spectra with a wave length of 360, 465, 495, 660, 733 and 800 nm have the greatest strength of connection..

The obtained regression equations can be used for remote research of the SCS of the Krasnoyarsk forest-steppe, subject to additional point studies carried out in a wider range of test parameters.

## Bibliography

1. Friedland, V.M. The structure of the soil cover. M .: Mysl, 1972 .-- 410 p.
2. Agroecological assessment of lands, design of adaptive landscape systems of farming and agricultural technologies. M .: Rosinformagroteh, 2005 .-- 783 p.
3. Pisman T.I., Shevyrnogov A.P., Lar'ko A.A., Botvich I.Yu., Emelyanov D.V., Shpedt A.A., Trubnikov Yu.N. Informative value of spectral vegetation indices for deciphering agricultural fields. Biophysics. 2019. No. 4. p. 740-746.
4. Shevyrnogov A.P., Botvich I.Yu., Emelyanov D.B., Lar'ko A.A., Vysotskaya GS, Ivchenko B.K., Demyanenko T.N. Possibility of recognizing the soil cover of the experimental field using ground and satellite data. Modern problems of remote sensing of the Earth from space. 2019.Vol. 16.No. 4.p. 150-160.
5. Map of Quaternary formations Sheet O-46-XXXIII scale 1: 200,000 second edition [Electronic resource] - Access mode: http://geolkarta.ru/list_200.php?idlist=O-46-XXXIII&idlist_d=Q&gen=2&g= 1 [Map of Quaternary Formations Sheet O-46-XXXIII scale 1: 200,000 second edition [Electronic resourse] - Access: http://geolkarta.ru/list_200.php?idlist=O-46-XXXIII&idlist_d=Q&gen=2&g=1 ]
6. Geological map Sheet O-46-XXXIII scale 1: 200,000 second edition [Electronic resource] - Access mode: http://geolkarta.ru/list_200.php?idlist=O-46-XXXIII&idlist_d=G&gen=2&g=1 [Geological map Sheet O-46-XXXIII scale 1: 200,000 second edition [Electronic resourse] - Access: http://geolkarta.ru/list_200.php?idlist=O-46-XXXIII&idlist_d=G&gen=2&g=1]
7. Weather archive in Sukhobuzimskoye [Electronic resourse] - Access: https://rp5.ru/Weather_archive_in_Sukhobuzimskoye]
8. Agrochemical methods of soil research. M .: Publishing house Nauka, 1965. -430 p.
9. Karavanova EI Optical properties of soil and their nature. M .: Publishing house of Moscow State University, 2003 .-- 153 p.

# DYNAMICS OF PRODUCTIVITY OF RESTORED VEGETATION OF THE IZYKHSKY SURFACE MINE (TERRA MODIS)

*Alexander* Zhukov[1*], *Elena* Zhukova[1]

[1] Graduate student, candidate of Biological Sciences, Federal State-Funded Educational Institution of Higher Education «Katanov State University», 655017, Lenin Avenue 90, Abakan, Russia

**Abstract.** The subject of the study is the course of the dynamics of vegetation productivity on the dumps of the surface mine «Izykhsky» (2000-2019), using Terra Modis. The aim is to identify patterns of seasonal and long-term dynamics of restored vegetation as a result of succession according to images. The methodology included processing information on gross primary production and evapotranspiration, as well as identifying the relationship with meteorological data for 2018-2019. There is a positive trend in the long-term dynamics – in terms of total gross primary production from 1.7 to 5.5 $kg/m^2/8$ days and in evapotranspiration from 1142 to 2784 $kg/m^2/8$ days. The evapotranspiration correlated with the productivity. Since 2016, productivity has reached a plateau, which indicates the development of ecological niches by plants. The phytomass of the restored vegetation was in 1.5 times greater, than the mass of the steppe site. Seasonal dynamics in 2019 showed that communities on dumps have higher productivity, in contrast to the steppe, and one peak in the first ten days of July. The sum of temperatures and productivity had a high relationship ($R^2 = 0.7$) in comparison with the steppe ($R^2 = 0.2$). Terra Modis data can be applied in the field of ecological monitoring of vegetation of coal dumps.

A monitoring of restored lands, the vegetation cover disturbances and the pollution is carried out also according to remote sensing data [1, 2]. The coal industry in Khakasia has increased dramatically since 2015 (7 surface mines, 5222 hectares of disturbed territories) [2]. Vegetation reconstruction efforts were becoming increasingly difficult in the areas affected by coal mining. The lack of moisture, high temperatures, with a short period of temperatures above 10° C leads to the fact that organic matter have started to mummify for a long time, and not its decomposition [3, 4].

The aim of the work is to identify patterns of seasonal and long-term dynamics of vegetation productivity as a result of succession after reconstruction of the Izykhsky surface mine (Khakasia) according to Terra Modis data.

For that purpose, it has been using Terra Modis data from the site https://lpdaacsvc.cr.usgs.gov/appeears: 1) gross primary production (GPP) [5], 2) evapotranspiration [6] (artifacts were replaced by the arithmetic mean from adjacent composites).

The sum of temperatures was obtained from the site rp5.ru. For integral productivity and evapotranspiration (half-sum of bases, multiplied by 8 days of the composite), the sum and standard deviation were calculated, and the regression was additionally calculated (Excel).

The Izykhsky coal surface mine has been developed since the 1960s. During this time, the natural landscape was transformed into the mining one in the form of a residual

* Corresponding author: zhukov.khsu@gmail.com

quarry and an external coal dump on the land surface of 131 and 322 hectares, respectively (Fig. 1). In the 2000s. its reclamation has begun [1].

The climate is continental, the radiation balance is 33-34 kcal per $cm^2$, water (65%) is spent on evaporation. The sum of temperatures for the warm season is 1800-1950°C. The amount of precipitation is 320 mm [3].

Chestnut soils, common or southern chernozems are widespread [3]. The layer up to 2 m is formed by Quaternary rocks. Humus in a layer up to 40 cm is 1.5–3.4% [1]. The evolution of steppe soils on coal dumps has been stopping for a long time only on the organo-accumulative stage, less often on the sod stage [7]. Due to the lamellar shape and angularity of embryozems, soil compaction occurs [4]. Erosion, deflation and aeolian weathering in the dry steppe zone can lead to a desertification and impeded a recovery [8].

There were *Artemisia glauca* Pall. ex Willd., *A. sieversiana* Willd., *Calamagrostis epigeios* (L.) Roth, *Cirsium setosum* (Willd.) Bess., *Elytrigia repens* (L.) Nevski, *Glycyrrhiza uralensis* Fisch. *Hieracium umbellatum* L., *Hordeum jubatum* L., *Kitagawia baicalensis* (I. Redowsky ex Will.) Pimenov, *Medicago falcata* L., *Melilotus albus* Medik., *Melilotus officinalis* (L.) Pall., *Poa pratensis* L., *Sonchus arvensis* L., etc. [1].

In 2020, ground-based studies of the shallow sod steppe (1 contour), vegetation on the southern (2), western (3) and northern (4) parts of the dump were carried out (Fig.1).

**Fig. 1.** Scheme of the Izykhsky coal dump: 1 – steppe (53.5943298° and 91.42101288°), 2 – southern part (53.60325623° and 91.43474579°), 3 – western part (53.61053467° and 91.43690872°), 4 – northern part (53.61527252° and 91.43182755°), 5 – eastern part (53.61808777° and 91.44054794°), 6 – central part of the dump (53.61080933° and 91.42228317°) (Google map, 1:200)

The steppe has a projective cover – 60%. The aspect of the steppe area is yellow-green with dove-colored spots of *Artemisia frigida* Willd. A simple two-level structure of the grass stand was noted with dominants in the first level – *Koeleria macrantha* (Ledeb.) Spreng. (30%) and in the second *Artemisia frigida* (20%) and *Potentilla acaulis* L. (5%). Of other species, there are *Artemisia scoparia* Waldst. et Kit., *Astragalus danicus* Retz.,

*Buglossoides arvensis* (L.) I.M. Johnston, *Chamaerhodos erecta* (L.) Bunge, *Elytrigia repens*, *Galatella angustissima* (Tausch) Novopokr., *Galium verum* L., *Koeleria glauca* (Spreng.) DC., *Leymus angustus* (Trin.) Pilg., *Melilotus officinalis*, *Onobrychis arenaria* (Kit. ex Willd.) DC., *Otites wolgensis* (Hornem.) Bess. ex Spreng., *Oxytropis pilosa* (L.) DC., *Poa angustifolia* L., *Potentilla longifolia* Willd. ex Schlecht., *Thymus minussinensis* Serg., *Veronica incana* L. etc.

In the ***southern part***, the projective cover is much higher (85%), the aspect is green, the structure of the herbage is three-level – *Ulmus pumila* L. (20%) prevails in the first layer, *Artemisia sieversiana* (7%), *Cirsium setosum* (7%), *Calamagrostis epigeios* (20%) and *Artemisia frigida* (10%). Other species included *Artemisia scoparia, Artemisia vulgaris* L., *Astragalus melilotoides* Pall., *Atriplex prostrata* Boucher ex DC. , *Bromopsis inermis* (Leyss.) Holub, *Calamagrostis epigeios, Cynoglossum officinale* L., *Festuca rubra* L., *Festuca valesiaca* Gaudin, *Glycyrrhiza uralensis, Hordeum jubatum, Lactuca sibirica* (L.) Benth. ex Maxim., *Lactuca sibirica, Lappula squarrosa* (Retz.) Dumort., *Lepidium latifolium* L., *Melilotus officinalis, Otites wolgensis, Poa angustifolia, Rumex crispus* L., *Spiraea media* Franz Schmidt, *Stipa capillata* L., etc. Spectral analysis shows that the southern and eastern slopes, as well as the western and central areas, have a similar species composition, but different projective cover.

The ***western part*** of the dump has a significant projective cover of 80%, and a variegated aspect (green with yellow spots of flowering *Melilotus officinalis*, purple flowers of *Cirsium setosum* and blue flowers of *Lactuca sibirica*. The structure of the community becomes more complicated up to 3 levels – *Ulmus pumila* dominates in the first level (10%, in places up to 20%), the second one is polydominant (*Calamagrostis epigeios* (20%), *Poa pratensis* (10%), *Cirsium setosum* (7%), *Melilotus officinalis* (5%)), in the lowest one there is a moss cover (30%). *Poa angustifolia, Erigeron acris* L., *Rumex crispus, Taraxacum bessarabicum* (Hornem.) Hand.-Mazz., *Sonchus oleraceus* L., *Melilotus albus, Hordeum jubatum, Artemisia sieversiana* are also found here.

In the ***northern part*** of the dump, trees and shrubs are well developed, in addition to *Ulmus pumila*, there are *Acer negundo* L., *Cotoneaster melanocarpus* Fisch. ex Blytt., *Populus laurifolia* Ledeb., *Padus avium* Mill., *Betula pendula* Roth, *Pinus sylvestris* L. The total projective cover is 65%, in some places up to 80%. Aspect – green with blue spots *Lactuca sibirica*. Two levels were recorded – *Ulmus pumila* (20%) prevails in the first, *Poa pratensis* (10%), *Cirsium setosum* (10%), *Calamagrostis epigeios* (10%) prevail in the second. Accompanying species include *Agrimonia pilosa* Ledeb., *Arctium leiospermum* (Juz.) et C. Serg., *Artemisia vulgaris, Centaurea stoebe* L., *Cynoglossum officinale* L., *Elytrigia repens, Erigeron acris, Festuca valesiaca, Galium verum* L., *Geranium sibiricum* L., *Geum aleppicum* Jacq., *Glycyrrhiza uralensis, Helictotrichon desertorum* (Less.) Nevski, *Hieracium umbellatum, Kitagawia baicalensis, Linaria acutiloba* Fisch. ex Rchb., *Poa angustifolia, Potentilla longifolia, Scabiosa ochroleuca, Sonchus arvensis, Stipa pennata* L., *Taraxacum bessarabicum, Vicia cracca* L.

From the point of view of studying of successions, the Terra Modis data are of interest primarily due to their significant archive (Fig. 2).

There are 3 peaks in 2003, 2014 and 2018, as well as declines in 2002, 2005, 2013 and 2015. The steppe has the smallest phytomass (*Koeleria macrantha, Artemisia frigida, Poa angustifolia, Potentilla acaulis*) and is characterized by stability in development, more resistant to unfavorable weather conditions, which is clearly marked in 2002 and 2005. There is a positive trend for all sites, especially for the eastern, northern and central parts.

High performance coincides with ground data. Evapotranspiration correlates with productivity (Fig. 3).

In the central part, evapotranspiration ranked first due to the accumulation of water (physical evaporation).

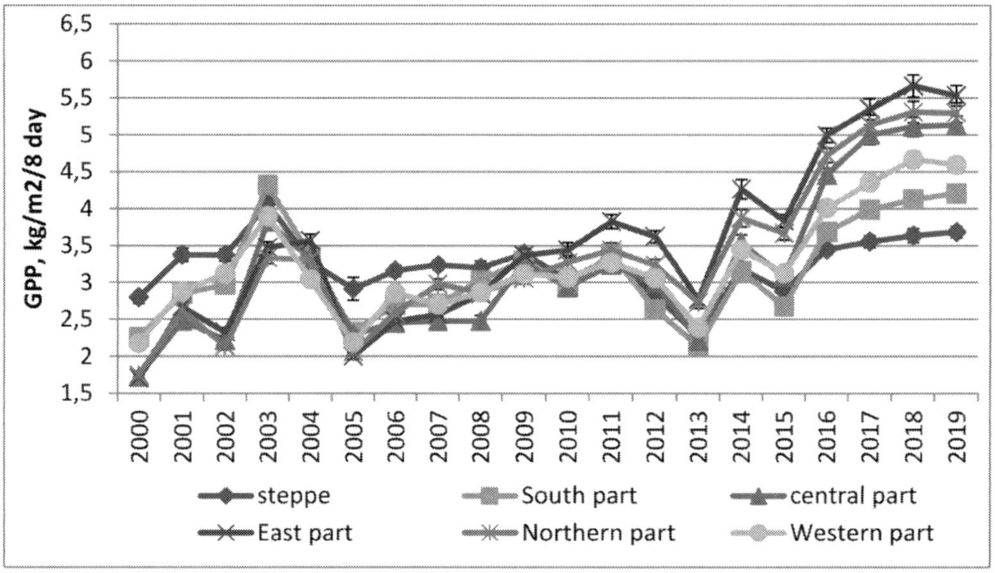


**Fig. 2.** The total gross production of the plant communities of the Izykhsky coal dump (2000-2019)

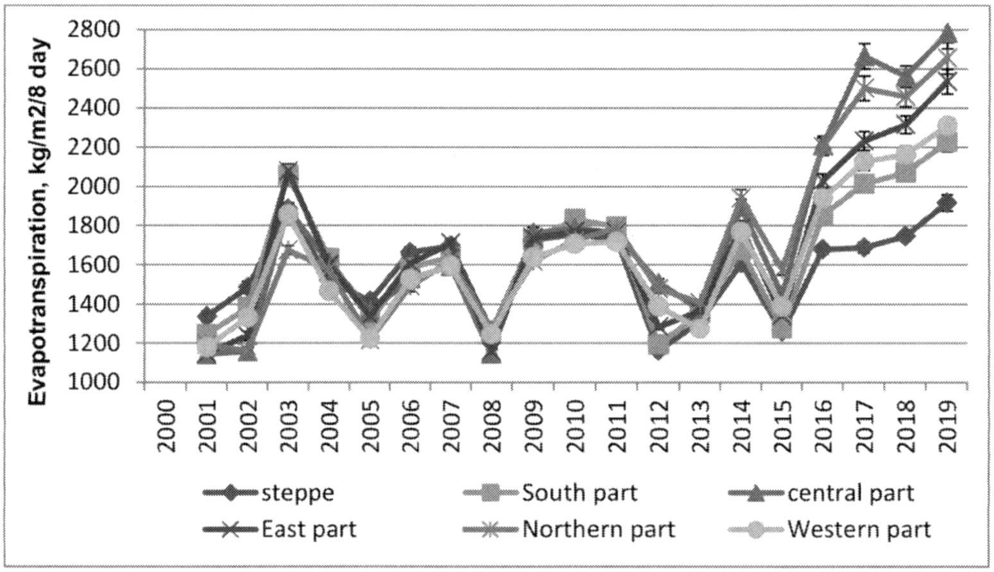


**Fig. 3.** The total evapotranspiration of plant communities of the Izykhsky coal dump in the period from 2000 to 2019

Consider the seasonal dynamics for 2019 (Fig. 4).

In contrast to the steppe with low productivity, the phytocoenoses on the dumps had a high productivity and a peak coinciding with the temperature peak. The maximum values were observed in the eastern and northern parts, the minimum values were in the south, where herbaceous vegetation is widespread.

The dependences of the GPP indicator on the sum of positive temperatures (regression) for the vegetative season from 2018 to 2020 were studied. The results showed that natural vegetation has a weak positive dependence on the sum of temperatures ($R^2$ = 0.16, 4 degree polynomial). This shows that the vegetation of such communities is better adapted from the point of ecological view to the semiarid climate.

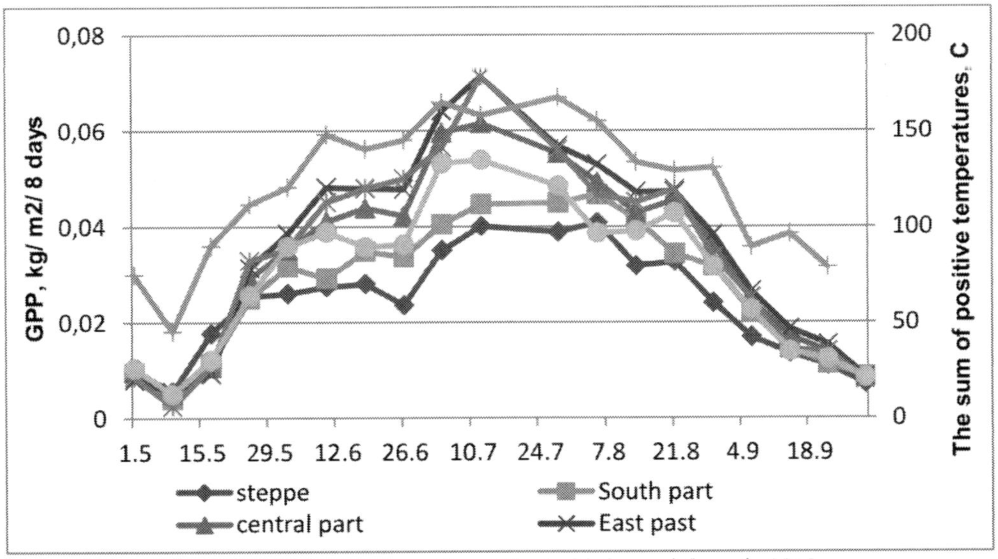


**Fig. 4.** Seasonal dynamics of plant communities of the Izykhsky coal dump in 2019

We list the indicators of the coefficient of determination for the polynomial equation of the 3rd degree – for the southern and western parts of 0.73 and 0.74, for the central, eastern and northern parts – 0.77, respectively; 0.75 and 0.77.

Consequently, the coal dump vegetation is more dependent on meteorological conditions than the steppe background

During the work, the following conclusions were made:

1. There is a positive trend – in terms of total gross primary production from 1.7 to 5.51.7 kg / $m^2$ / 8 days and for evapotranspiration from 1142 to 2784 kg / $m^2$ / 8 days in the long-term dynamics of Terra Modis satellite data in the period from 2000 (2001) to 2019. The dynamics of evapotranspiration correlates with gross productivity in general. Since 2016, the GPP index has reached a plateau, which indicates the closure of the vegetation cover and the occupation of accessible ecological niches by plants.
2. The GPP parameter for the restored vegetation was 1.5 times higher than the natural steppe vegetation. Consequently, for plants on the coal dump, in general, the growing conditions (moisture, nutrients) are favorable. This is confirmed by a higher projective cover and larger sizes of species. This makes the territory promising for the restoration of biological diversity, even for trees and shrubs.
3. The analysis of seasonal dynamics for 2019 showed that, in contrast to the steppe community, the communities on the restored coal dumps had high productivity (up to 0.07 kg / $m^2$ / 8 days) and one peak of the seasonal development in the first ten days of July. The maximum values throughout the season were observed in the more humid eastern and northern parts of the coal dump, and the minimum values were in the southern part.
4. Regression analysis showed that GPP and the sum of positive temperatures had a high level of relationship ($R^2 = 0.7$) compared to the steppe site ($R^2 = 0.2$).

## References

1. O.A. Berezina, A.N. Shikhov, R.K. Abdullin The use of multi-temporal satellite images for environmental assessment in coal mining areas (by example of closed Kizel coal basin) // Current problems in remote sensing of the Earth from space, **15 (2),** 144-158 (2018)

2. L.V. Zenkov, B.N. Nefedov, Yu. P. Yuronen, N.B. Nefedov Environmental condition remote soundining and field surveys of the lands, disturbed by coal mining open pits in the republic of Khakassia //Coal, **9,** 72–75 (2017)
3. Kuminova A.V. et al. Vegetation cover of Khakasia. Novosibirsk: Science, Siberian Branch RAS,127 p. (1976)
4. E.A. Gurkova, D.A. Sokolov, S.P. Kulizhsky, S.V. Loiko Micromorphological peculiarities of soils in technogene landscapes of coal-mine industries in Siberia (Russia) // Novel methods and results of landscape research in Europe, Central Asia and Siberia, 27–31 (2018)
5. Running S., Mu Q., Zhao M., MOD17A2H. doi: 10.5067 (2015) (date 25.11.2018).
6. Running S., Mu Q., Zhao M. (2017). Terra Net Evapotranspiration. MOD16A2.006. doi: 10.5067 (date 25.11.2018).
7. D.A. Sokolov, S.P. Kulizhsky, S.V. Loiko, E.A. Domozhakova Using electronic scanning microscopy for diagnostics of soil-forming processes on the surface of coal-mine dumps in Siberia// Tomsk State University Journal of Biology, **3,** 36–52 (2014)
8. Balyazin I.V. Influence of the coal industry on the ecological condition of the Koybalskaya steppe // Geographical Research of Asian Russia and Adjacent Territories: New Methods and Approaches, 28–32 (2019)

# THE MONITORING SYSTEM OF HAZARDOUS TECHNOLOGICAL OBJECTS BASED ON UNMANNED AERIAL VEHICLES

*Ivan* Zoev[1*], *Nikolai* Markov[2], and *Svetlana* Ryzhova[1]

[1]National Research Tomsk Polytechnic University, Tomsk, Russia
[2]Dr.Tech.Sci, National Research Tomsk Polytechnic University, Tomsk, Russia

**Abstract.** The paper describes the system for monitoring of hazardous technological objects, which based on unmanned aerial vehicles. The feature of the system is the using of the intelligent computer vision subsystem in which hardware-implemented a convolutional neural network based on a system-on-a-chip. This subsystem recognizes objects of the earth's surface and made decision about incidents at the monitoring area in real time. It also has service-oriented software architecture for ground-based system components. The paper presents results of the research of the required resources of a field-programmable gate array for implementation of error-correction encoder-decode based-on-board.

## 1 Introduction

Nowadays in Russia and many foreign industrialized countries have problems of monitoring the situations and detecting illegal actions at hazardous technological facilities. Also it has a problem of timely detection of fires at such facilities and in the territories they belong. Solve of these problems is primarily relevant for such industries, that have complex technological processes and facilities. The more popularity becomes unmanned aerial vehicles (UAVs) with special equipment for monitoring such facilities installed on them. Usually on a board UAVs are installed a video camera and thermal imager, which allows to shooting the monitoring territory. The existing monitoring systems don't analyze on the UAVs obtained images. These images are transmitted via radio-channel to the ground part of system and then operator analyzes (interprets) them. The lack of images real-time analysis on the UAV doesn't allow timely to make a decision in the ground part of monitoring system and send an alarming group to the place of the incident. As a result, all of this does not allow understanding the technical condition or prevent an invasion in a technological facility.

Earlier in our article [1] we proposed an idea of images analysis during the monitoring of earth's surface on a board UAVs in real-time (at the images receiving rate from a video camera and thermal imager). To realization of this idea we proposed to include an intelligent computer vision subsystem (ICVS) to the monitoring system and installed it on

* Corresponding author: zoev.ivan@yandex.ru

the UAV. This subsystem will allow analyzing obtained data in automatic mode and made a decision about an incident on monitoring territory.

In this article we describe a concept of developed monitoring system of hazardous technological facilities based on modern UAVs and ICVS. This monitoring system may be autonomous or may be included in center of territories monitoring control. Also we will present features of its subsystems.

## 2 Concept of monitoring system

The system includes an UAV of aircraft-type or helicopter-type, ICVS installed on UAVs, ground-based information monitoring control subsystem and data transmission subsystem. Last of these subsystem allows transmit via radio-channel instructions to UAV and receive a monitoring data from its.

Choosing of UAVs type depends on technological facilities features to be monitored and the tasks, which will be solved using the system. UAV must include flight controller, GPS module, battery designed for significant flight time, engine(s) with speed controller and etc. Note that helicopter-type UAV can be located above the monitoring facilities and had power from the ground, because of its stationary location.

To monitor large areas, which includes several hazardous facilities (magistral oil and gas pipeline and etc.) may be developed a monitor control center. In this case information monitoring subsystem may be a part of this center and monitoring subsystem will not be autonomous. Operators of this control center should be able of quick interpretation of obtained from UAV images, if it necessary in doubt about an incident to made a final decision about sending of alarming group. Situation like this takes a place, because not all incidents may be identified by ICVS correctly. All these mean that the information subsystem should be integrated with one or several subsystems of automatic interpretation of obtained images in the center.

## 3 Intelligent computer vision subsystem

ICVS installed on UAV and included photo or video camera, thermal imager for detection of fire-sources and computing unite (CU). As a CU core we used a modern system on a chip (SoC). CU manages the collection of earth's surface objects images, analyze them using convolutional neural network (CNN) and automatic decides about the presence of incidents in monitoring areas. The main requirement to ICVS is the performance of all these function in real-time. According the high requirements from CNN to CU performance, we proposed a hardware-based implementation of promising CNN architectures on Field Programmable Gate Array (FPGA) SoC.

We have implemented two variants of ICVS. First of these variants is hardware-implemented CNN of original architecture from LeNet5 class on Cyclone V SX SoCs [2]. This CNN solving task of classification of objects of various physical natures (technological objects, special equipment, people and etc.) in images. The second variants of ICVS has CU with SoC Zynq 7000 (Kintex FPGA) as a core. On the FPGA of this Soc we hardware implemented an original tiny-YOLO-InceptionResNet2 CNN architecture of YOLO class [3].

We researched every of these hardware-implemented CNNs using dataset called "Stanford Drone Dataset" [4]. Results of these researchers helped to select of CNN hyperparameters in this way that to recognize objects in images with practically acceptable accuracy. Moreover each ICVS variants may function in real-time.

## 4 Information monitor control subsystem

The information subsystem has a software (SW) of well-known freely geoinformation system (GIS) QGIS as a core [5]. Operators, who work with this subsystem, can plot and track UAVs routes, display the coordinates of incident (objects) in electronic map and etc using this SW. The information subsystem may be included in both an autonomous monitoring system and a monitoring control center with monitoring system.
We based on service-oriented architecture (SOA) principles when developed software for information subsystem and developed a unified information space of control center. Considering this, we developed SW center structure (fig.1). In figure 1 colored components, which belongs to information subsystem. The basis of SOA model are ELMA BMP business-processes control system and enterprise service bus ESB. They allow cooperating subsystem with third-party services, with database (through a database management system – DBSM) and with other systems automated interpretation systems of center, for example with ERDAS system. Figure 1 shows third-party web-services of error-correcting data encoding-decoding, included in ground-based part of data transmission subsystem. We have developed adapters for systems integration with ESB-bus.
We added an "Active UAV" module into standard SW QGIS, developed in its environment.

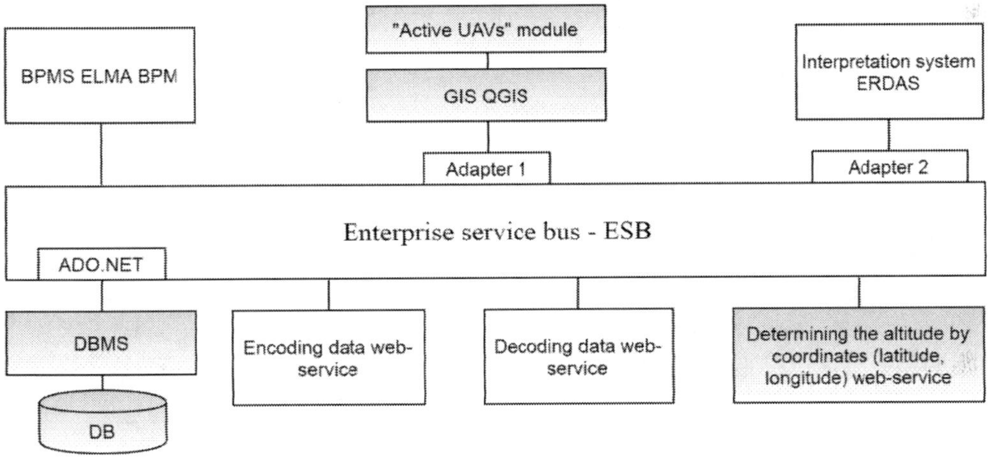


**Fig. 1.** Structure of SW of monitoring control center

The main module window is marked by rectangle in figure 2. An ovals in this figure presents enlarged functions, which performed by module (the dotted line indicates, that function is running in background). And the hexagon shows QGIS core functions.
This module's functions of UAVs adding/removing, UAVs replacing and viewing of received images must connect with MS SQL Server 2017 DBMS, which allows maintaining the database (BD). This BD stores information about UAV, images transmitted from UAV to monitoring control center, and other service information. The function of tracking UAV in flight designed to route in real-time the UAV's flight on the map. This function has tight connection with QGIS core, which provides all necessary tools for displaying the route in the map.

https://doi.org/10.1051/e3sconf/202022303005

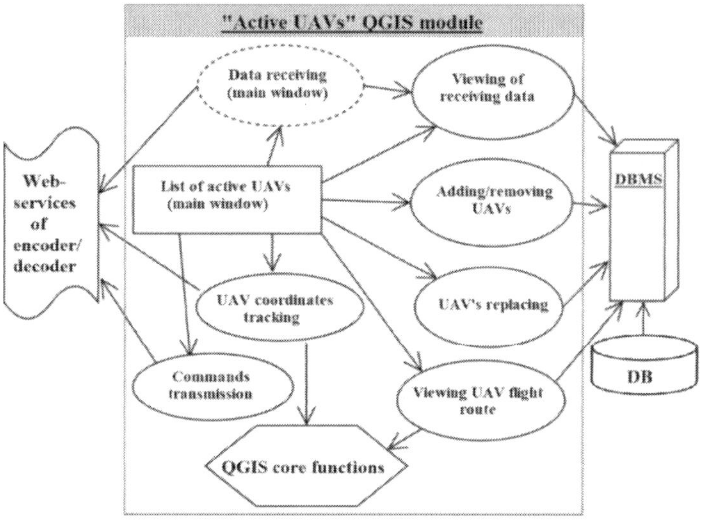


**Fig. 2.** The relationship diagram between module elements "Active UAVs"

Commands transmission to the UAV and data receiving from the UAV are carried out through web-services. These web-services carry out an error-correcting encoding and decoding of all transmitted and received information.

We have developed and have software implemented algorithms of cooperation of "Active UAVs" module with ICVS using C# language. For developing an algorithm of transmission an image from UAV to information monitoring control subsystem we have took into account that the implementation of image viewing function requires the cooperation of the QGIS with DB. This cooperation was developed using SQL Python-pyodbc driver and QGIS PythonAPI [5, 6].

## 5 Data transmission subsystem

The exchange of commands and data between the onboard and ground-based parts of the data transmission subsystem is carried oud via a radio-channel. The onboard part in UAV has a radio-module and the ground-based part has a transceiver for such exchanges. To increasing of reliability of transmitted commands and data the onboard part and the ground-based part of data transmission subsystem can encode these data using error-correcting codes for detection and correction of errors in them. Among error-detection codes we have chosen a cyclic CRC code. Researches showed that the best option for CRC computation on FPGA is a matrix algorithm.

Encoder in the UAV make error-correction data encoding, which will be transmitted to the ground-based part of subsystem, using polynomial cyclic encoding algorithm. Figure 3 presents researches results of the Altera amounts FPGA resources requires by encoder depending from the data length n, that counts in bits on the SoC Cyclone V SX.

We have researched lengths of data $n = 31, 63, 127$ and 255 bits. The amount of necessary FPGA resources was determined by the number of required ALMs-blocks (adaptive logic modules) and ALUTs-blocks (adaptive look-up table). The analysis showed that for data length $n \leq 127$ bits the amounts of needed FPGA resources for encoder implementation is less than we can allocate for this encoder. Similar results of researches we have obtained for FPGA-based decoder, which using a cyclic decoding algorithm.

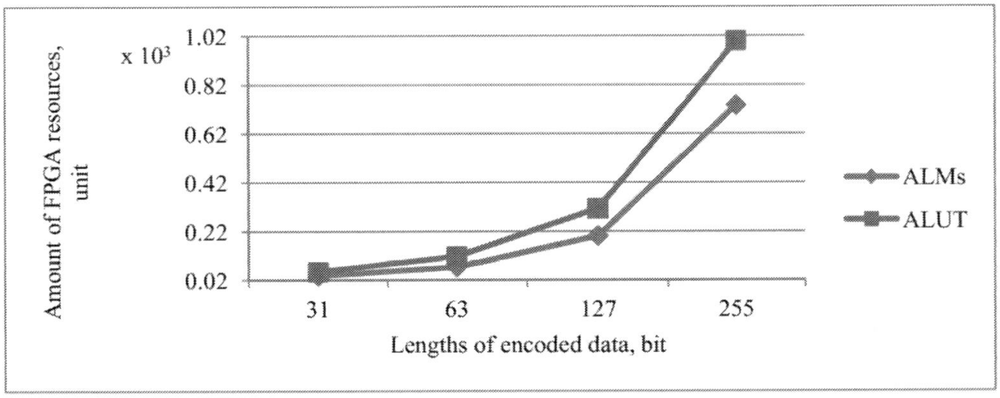


**Fig. 3.** Dependence of amounts of FPGA resources needed for ecnoder from length of encoding data

The ground-based part of data transmission subsystem has software-implemented of error-correcting encoder-decoder as web-services (fig.1) using C# language.
We have researched the decoding speed of receiving form the UAV data depending on the encoded data lengths (n) and number of errors using table and cyclic decoding algorithms. Researches were carried out for 50 sending, each of them containing of 50 Kbit of data, and for number of errors from 1 to 7 bits in a row. We may a conclusion based on this researches that decoding time using table algorithm increases insignificantly with an increasing of errors number in encoded data. But for cyclic algorithm in this case decoding time increases at 1.5 times. However, the table algorithm requires a significant memory amount for table storing. Moreover, this memory amount depends on data lengths and error correcting capability, and increases exponentially with the lengths. As a result, we have chosen a cyclic decoding algorithm for implementation. Table 1 shows results of researches decoding time using this algorithm. This table contains minimum, maximum and mean time, necessary for decoding of data various lengths (Tmin, Tmax and Tmean respectively).

**Table 1.** Result of researching of decoding data time using cyclic algorithm in ground-based part of transmission subsystem

| n | $T_{min}$, sec | $T_{max}$, sec | $T_{mean}$, sec |
|---|---|---|---|
| 31 | 9.088 | 9.126 | 9.101 |
| 63 | 9.984 | 10.210 | 10.091 |
| 127 | 20.754 | 20.895 | 20.812 |
| 255 | 66.249 | 67.206 | 66.618 |

These obtained results indicate that the optimal data lengths, received and decoded in ground-based data transmission subsystem part, is n=127 bits. For data with bigger length we have noticed a sharp increasing of time, needed for decoding, which doesn't fulfill the requirements of real-time monitoring system operation.

## 6 Conclusion

Nowadays the development of mobile and fundamentally new monitoring systems for hazardous technological facilities is important. We have developed the concept of such system based on UAV and have implemented its prototype. Also, we have tested this systems prototype in laboratory conditions and results of these test indicates about the efficiency of all its subsystems.

https://doi.org/10.1051/e3sconf/202022303005

## Acknowledgment

The research was supported by Russian Foundation for Basis Research grant No. 18-47-700010 r_a.

## References

1. I.V. Zoev, N.G. Markov, S.E. Ryzhova Intelligent computer vision system for unmanned aerial vehicles for monitoring technological objects of oil and gas industry, Bulletin of Tomsk Polytechnic Uni. Geo Assets Eng, **330**(*11*), (2019)
2. I.V. Zoev, A.P. Beresnev, N.G. Markov, A.N. Malchukov FPGA-based device for handwritten digit recognition in image, Computer Optics, **41**(*6*) (2017)
3. I.V. Zoev, A.P. Beresnev, N.G. Markov Convolutional neural networks of the YOLO class in computer vision systems for mobile robotic complexes, Proc. IEEE Inter. Sib. Conf on Control and Comm., (2019) URL: https://ieeexplore.ieee.org/stamp/stamp.jsp?tp=&arnumber=8729605
4. Stanford Drone Dataset: Multi-scale, Multi-target social navigation, URL: https://stanford.resoluteinnovation.com/technologies/S17-376
5. Official site of QGIS, URL: https://www.qgis.org/ru/site/
6. QGIS 3 Plugins – Qt Designer for Plugins, URL: https://gis-ops.com/qgis-3-qt-designer-explained

# Adaptation to climate change in Kazakhstan using data from space monitoring of snowmelt

*Aknur* Zholdasbek [1], and *Azamat* Kauazov[2,*]

[1]Al-Farabi Kazakh national university, faculty of geography and environmental sciences, Almaty, Republic of Kazakhstan
[2]RSE "KazHydroMet" (National weather service), Research Department, Nur-Sultan, Republic of Kazakhstan

**Abstract.** The present article is concerned with the applied aspects of applying the results of space monitoring of snow cover, in particular, it is proposed to present the results of the analysis in the form of specialized bulletins. The purpose of this publication is to present the available results of space monitoring of snow cover in Kazakhstan as an element of adaptation to climate change. A three-level system of space monitoring of snow cover is proposed, which includes three technological complexes: operational mapping of snow cover boundaries; monitoring of seasonal snow cover dynamics; analysis of long-term snow cover dynamics. A map of snow melting in Kazakhstan in 2020, the dynamics of long-term changes of snow covered area, as well as methods for analyzing the spatial-temporal distribution of snow cover and formats of special bulletins are presented. It is most appropriate to present the results of space monitoring of snow cover in a complex, maximally generalized form (product). The results of the work can be applied in the scientific, industrial and educational spheres to adapt and increase resistance.

## 1 Introduction

Snow cover plays a significant role in the climate system, in emergencies, and in agriculture. The range of applications of snow cover data is extremely broad. Initially, the most accessible task for remote sensing from space was to monitor snowmelt. Regular space monitoring of snowmelt has been carried out in Kazakhstan since 2000. Certain results have been achieved in the field of seasonal and long-term monitoring of snow cover according to the remote sensing data [1;5]. In practical terms, the results of snowmelt monitoring can be used in important applications: assessment of global climate change, to control the passage of spring floods in regions with a high risk of flooding, assessment of agricultural risks taking into account the impact of weather conditions (assessment of productive moisture reserves in the soil, assessment of the probability of dry conditions). The last appendix is of particular importance for non-irrigation grain cultivation areas in the Northern regions of Kazakhstan, since spring soil moisture is a key factor determining the yield of grain crops. A number of studies [6,7] have shown the possibility of forecasting

* Corresponding author: kauazov_a@meteo.kz

droughts based on the height and position of the snow cover boundary in the spring period. The author [7] found that severe droughts are observed after the early snow melting. On the contrary, when there was a late snowmelt (the snow cover borders are South of the usual one), there was no drought, and the late snowmelt reduce the risk a moisture shortage in the spring-summer period. In Kazakhstan, the use of long-term remote sensing data and the basics of creating space monitoring systems are most well covered in the work [8].

In general, in the context of climate change in Kazakhstan, it is expected to reduce the of snow covered area (SCA) and its earlier descent. Snow cover is particularly sensitive to temperature changes in the periphery, in the zero temperature zone. Kazakhstan belongs to the border zone, southern Kazakhstan is an area with unstable snow cover, in Northern Kazakhstan, snow still persists for 4-5 months a year and is subject to significant risks of reduction of SCA and early descent. Taking into consideration that Northern Kazakhstan provides 80 % of the spring wheat yield, informing farmers becomes an urgent task. In addition, adaptation strategies recommend improved monitoring and awareness of farmers. Therefore, informing farmers about the snow cover is a win-win adaptation, since the best informing of consumers is necessary by itself, regardless of conditions.

The main purpose of this publication is to briefly describe the structure of snow cover monitoring in Kazakhstan and demonstrate the preliminary concept of the format for providing data to consumers, as a development of the ideas proposed by L. F. Spivak [8].

## 2 Results and discussion

### 2.1 Recognition and operational mapping of snow cover borders

The Normalized Differential Snow Index (NDSI) is used for snow cover recognition, which takes into account the high reflectivity of snow in the visible range and the low emissivity in the near-infrared (IR) range [9]. Currently, the vector borders of the snow cover are available daily for download on the website worldview.earthdata.nasa.gov 1-2 times a day based on Aqua and Terra MODIS data with a delay of 1 day. This time lag is not critical for the tasks of this monitoring. Surely, these NDSI products contain inaccurate and erroneous areas occupied by snow cover (a typical example is how salt from the bottom of the Aral sea is classified as snow), but in generally, the border of the snow cover is distinguished with acceptable accuracy. In General, this task is evaluated as solved.

The problem of assessing changes is solve by daily monitoring of SCA, which allows you to assess the dynamics of snow cover loss.

### 2.2 Monitoring the seasonal dynamics of snowmelt

The technology for analyzing the seasonal dynamics of snowmelt provides for spatial comparison of operational maps and assessment of changes in SCA, snow-free territories, and active snowmelt zones. It should be noted that the snowmelt is uneven. The comparison results are drawn up in the form of maps and diagrams that reflect changes in the area of zones, the analysis of which allows you to estimate the rate of snowmelt and identify areas of accelerated snowmelt. This information is useful for predicting the most critical areas of flood development and can be used to estimate the amount of snow that has melted.

At the end of the snowmelt process, a final map is formed (Figure 1) on the entire territory of Kazakhstan by combining daily vector data on the boundaries of the snow cover, which is then used to analyze the inter-seasonal dynamics of snowmelt.

In General, at this level, based on the results of the analysis of snowmelt, it is advisable to form a Bulletin once a decade that reflects the sub-decadal dynamics of snowmelt in the current season. The optimal periodicity of the Bulletin is 1 time per decade according to the "sliding window" principle.

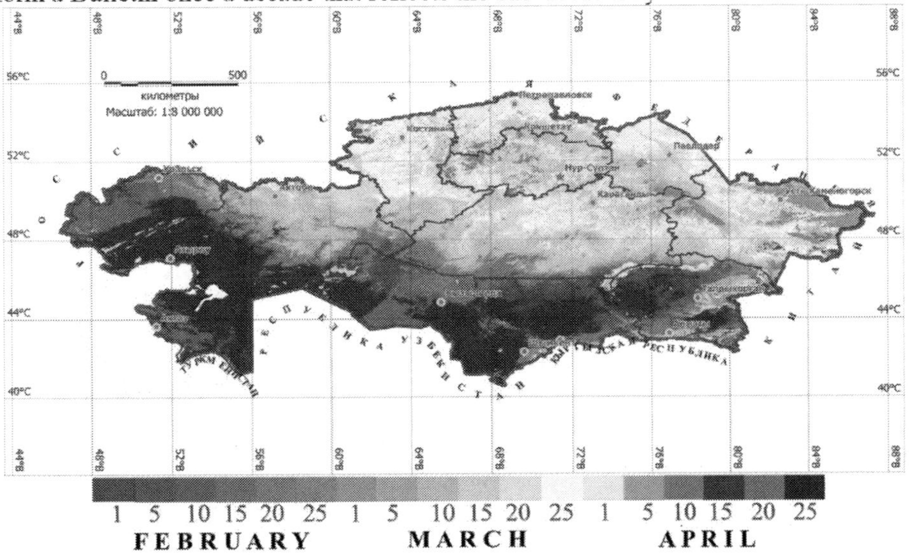


**Fig. 1.** Snow melting map for 2020.

## 2.3 Analysis of long-term dynamics of snowmelt

The SCA obtained for a long-term period allow the formation of long-term data series. Figure 2 shows the dynamics of changes of SCA obtained from MODIS data. There is a trend towards a reduction of SCA. The most likely cause is global warming. Long-term data series allow statistical comparisons of different seasons.

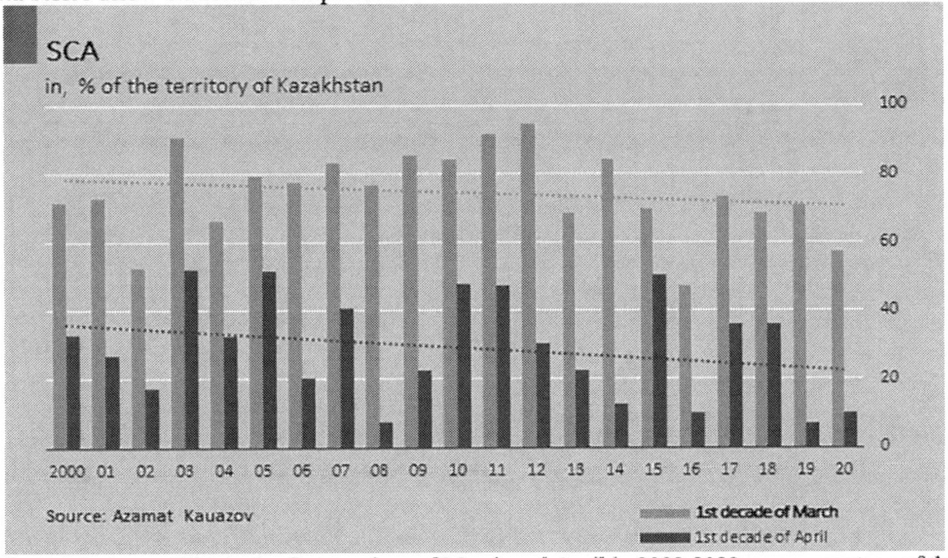


**Fig. 2.** Maximum SCA in the first ten days of March and April in 2000-2020, as a percentage of the territory of Kazakhstan.

Off-season monitoring includes:

- comparative analysis of snow cover dynamics in the current and previous seasons to assess the risk of flooding of territories;
- analysis of spatio-temporal patterns of snow cover loss and development of droughts over a multi-year period.

A comparative analysis of the results of monitoring snowmelt in the current season and in previous years allows us to estimate the rate of snowmelt relative to the average annual norm (fast, normal, slow) and select a "year analog" similar to the rate of snowmelt, which makes it possible to predict the development of the situation in the current year according to a similar scenario.

The results of the analysis are issued in the form of a special bulletin that provides an opportunity to compare current conditions with previous years. The consumer should have a visual opportunity to assess the differences in rate of snow melting, find out how sooner or later the snow melts, and accordingly prepare for sowing.

It is equally important to be able to objectively compare the dynamics of snowmelt in the current season with long-term observations, choose the most suitable "years analog" and, taking this into account, specify the sowing dates.

Figure 3 shows an approximate concept design of the bulletin as a result of comparing the dates of snowmelt in the current season with the average annual norm, indicating the zones of early and late snowmelt.

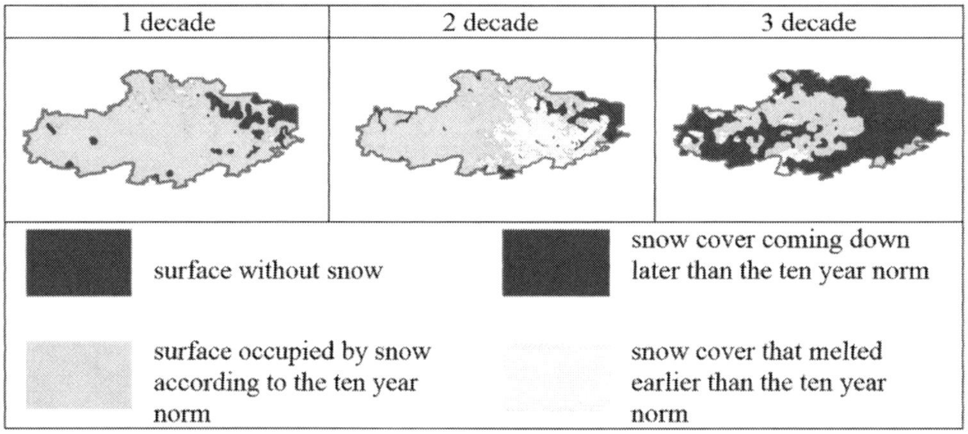


**Fig. 3.** Analysis of the dynamics of snowmelt with a ten-year norm.

Monitoring the dates and rates of snowmelt is also important because it allows you to indirectly predict potential droughts. It is known that the first half of the vegetation period of plants largely depends on snow reserves, or rather on the spring moisture reserve in the soil. Comparison of vegetation maps (NDVI) and snow cover maps shows that areas with low snowmelt are potential sources of drought events [10]. Early snowmelt and, accordingly, early spring, cause an early start of vegetation, while the maximum vegetation also shifts to the early dates: end of may-beginning of June, although in normal years the peak vegetation occurs a month later. Rapid development of vegetation, coupled with low initial soil moisture reserves, causes a lack of moisture and the subsequent development of drought.

It is obvious that consumers are interested in the final product, the moisture content in the snow cover, or the soil moisture content. Snow bulletins are planned to be implemented on the basis of the national hydrometeorological service of Kazakhstan RSE

«Kazhydromet». The proposed approach for estimating the area of snow cover is only an initial stage. The objective complexity of the snow cover, the lack of necessary satellite data/sensors , as well as the lack of models that accurately describe all processes in the snow cover does not allow using data on moisture storage with sufficient accuracy and reliability-figure 3. The difficulties and prospects of remote snow water equivalent (SWE) assessment are described in more detail in the SnowEx project.

Some hopes in the near future are associated with the launch of a Chinese satellite as part of the WCOM project. FEWS NET data is also promising if it is calibrated. A comparative analysis of FEWS NET SWE data and ground-based observations has shown that there are significant absolute and relative errors.

Besides, Kazhydromet is working on the adaptation and calibration of hydrological and numerical models, for example, the MODSNOW model is being improved and the SNOW-17 model can be adapted in the future.

Thus, despite some difficulties in monitoring snow cover at the global level, at the regional level, there is hope for a significant improvement in the snow cover monitoring system, increased preparedness and adaptation.

## 3 Conclusion

Monitoring is not an end in itself. The data set must be analyzed to get new information. At the same time, it is important not only to analyze and summarize the monitoring results, but also to present them in a convenient and visual form. Since data on snow cover are very diverse in the spatial and temporal aspect and may differ significantly from year to year, therefore, it is most appropriate to present the results of space monitoring of snow cover in a comprehensive, as generalized as possible, which can help attract a wide range of consumers to the monitoring results. Against the background of a reduction of snow covered area, the proposed three-level system of space monitoring of snow cover seems to be a logical and integral result for the formation of a full-fledged final product. In addition, awareness-raising activities for farmers and the public are an important area of climate change adaptation and capacity-building.

## References

1. A.M. Kauazov, A.C. Dara at all. Sovremennye problemy distantsionnogo zondirovaniya Zemli iz kosmosa (Sovr. Probl. DZZ Kosm.), **13. № 1**, p. 161–168, (2016) DOI: 10.21046/2070-7401-2016-13-1-161-168
2. A. G. Terekhov, N. I. Ivkina, N. R. Yunicheva, I. S. Vitkovskaya, A. G. Eltay, Sovremennye problemy distantsionnogo zondirovaniya Zemli iz kosmosa (Sovr. Probl. DZZ Kosm.), **17, № 2,** p. 101-113, (2020) DOI: 10.21046/2070-7401-2020-17-3-31-40.
3. A.G. Terekhov, N.N. Abaev, N.R. Unicheva, Sovremennye problemy distantsionnogo zondirovaniya Zemli iz kosmosa (Sovr. Probl. DZZ Kosm.), **16, № 5,** p. 351–355, (2019) DOI: 10.21046/2070-7401-2019-16-5-351-355.
4. V. G. Salnikov, G. K. Turulina, E. A. Talanov, S. E. Polyakova, A. M. Kauazov, V. V. Vorotyntseva, Proceedings of the Hydrometeorological Research Center of the Russian Federation, **358,** p 133-144, (2015)
5. A. G. Terekhov, Gidrometeorologiya i ekologiya, **90, № 3**, p. 29-36, (2018)
6. M. A. Lokoshchenko, Meteorologiya i Gidrologiya, **6**, p. 71-82, (2005)
7. V. V. Popova, Meteorologiya i Gidrologiya, 8, p.78-90, (2004)

8. L. F. Spivak, Methodological guide: Fundamentals of creating space monitoring systems, **2**, p.88, (2010)
9. I. A. Neustadt, Sovremennye problemy distantsionnogo zondirovaniya Zemli iz kosmosa (Sovr. Probl. DZZ Kosm.), **3,** p.359-365, (2006)
10. E.A. Zakarin, L. F. Spivak, Remote sensing methods in agriculture in Kazakhstan, p.230 (1992)

# Remote sensing data for assessing the equivalent water height (EWH) variety versus the level of forest disturbance in Central Siberia

*Konstantin* Krasnoshchekov[1,*], *Alexander* Dergunov[1] , and *Evgenii* Ponomarev[1, 2]

[1] Federal Research Center Krasnoyarsk Science Center of the SB RAS, Krasnoyarsk, Russia
[2] Siberian Federal University, Krasnoyarsk, Russia

**Abstract.** The forests of the permafrost zone of Central Siberia are influenced by natural and anthropogenic factors. Using remote sensing data, disturbed areas of forest cover are detected in the IR range as an anomalies of the surface temperature, which are significant for a long time (~20 years). Long-term changes in the temperature balance of the surface affect both the state of the lower soil horizons and the dynamics of the seasonally thawed layer, and, therefore, can affect the variation in the water balance and groundwater runoff. For the area of interest, a trend is shown of ~20% increase in the total area of disturbances per last two decades. A correlation analysis between large-scale thermal anomalies of the underlying surface and the series of gravimetric data on Equivalent Water Height (EWH) parameter is performed for the territory of the two river basins of the Nizhnyaya Tunguska and Podkamennaya Tunguska (Central Siberia). As a first approximation we estimated the relationship between the forest and on-ground cover disturbance and thermal anomalies of the underlying surface, which affect the seasonal dynamics of groundwater.

## 1 Introduction

The forests of Central Siberia are subject to the complex impact of various anthropogenic, technogenic and natural factors. The development of the northern territories, logging activities, mining operations lead to an increase in the area of disturbed forest litter. At the same time, wildfires are remain one of the main factors of damage to vegetation, which account for up to 70% of the total number of fires in Eurasia [1].

Using remote sensing data, it was found that forest areas affected by various destructive factors, due to changes in spectral characteristics in a wide range of the spectrum, are confidently detected in the thermal IR range in the format of thermal anomalies of the underlying surface from satellites [2–4]. Such anomalies, as it was shown in a number of recent papers [5, 6], can cause changes in the stability of ecosystems in the region.

Also it has been shown [2, 6–9] that large-scale disturbances of vegetation and the anomalies of the temperature balance of the underlying surface (thermal anomalies) caused

* Corresponding author: krasko@icm.krasn.ru

by wildfires affect the state of the permafrost layer, the level of groundwater and the dynamics of river flows. In order to verify this statement, we used a product of the GRACE satellite system, which provides gravimetric information expressed in terms of the equivalent water height (EWH).

The purpose of this work is a geospatial analysis of the degree of destructive impact of natural and technogenic factors in the permafrost zone of Central Siberia. As well as, the assessment of the influence of surface thermal anomalies on the groundwater runoff within river basins of the study area.

The study area was the permafrost zone of Central Siberia (56.8° – 68.3° N and 87.2° – 114.9° E). We used data for the basins of two rivers: Nizhnyaya Tunguska with an area of 473,000 km$^2$ and Podkamennaya Tunguska with an area of 240,000 km$^2$. These basins are presented in a scale of the 4th level (of 12$^{th}$ available) according to the classification Hydrological data and maps based on Shuttle Elevation Derivatives at multiple Scales (HydroSHEDS) (Fig. 1) [10].

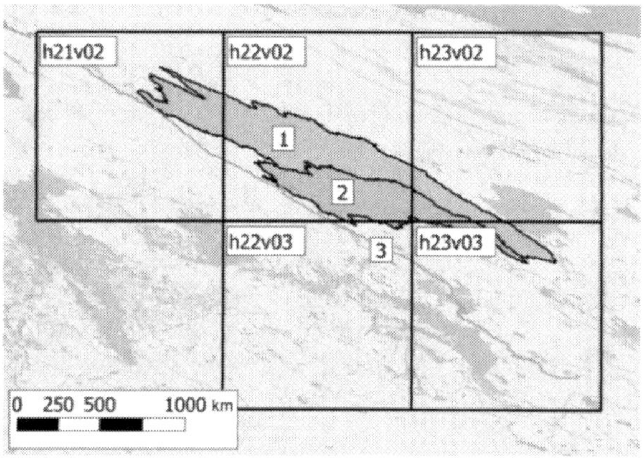


**Fig. 1.** Image of the study area and five MODIS grid cells in the WGS84 projection. Number 1 denotes the basin of the Nizhnyaya Tunguska river, number 2 – the basin of the Podkamennaya Tunguska river, number 3 - the Yenisei river.

For the study area, an archive of data on the average monthly surface temperature in the summer seasons (product MOD11B3) for the period from 2001 to 2019 and gravimetric data expressed by the EWH parameter for the period from 2013 to 2019 was formed.

## 2 Materials and methods

### 2.1 Data of the temperature of the underlying surface

To analyze the temperature of the underlying surface within the boundaries of the study area, the data of the IR range of the MODIS instrument installed on the Terra spacecraft were used. This is a MOD11B3 product, which is a daytime composite monthly average images using 20, 22, 23, 29, 31 and 32 channels of the MODIS radiometer with a spatial resolution of 6 km. The data was obtained from the LAADS DAAC resource (NASA) https://ladsweb.modaps.eosdis.nasa.gov/ [11].

The surface temperature maps of the study area were obtained on the MOD11B3 product within the boundaries of river basins (Fig. 2). The data sample consisted of 5 images corresponding to the following MODIS grid numbers: h21v02, h22v02, h23v02, h22v03, h23v03 (Fig. 1).

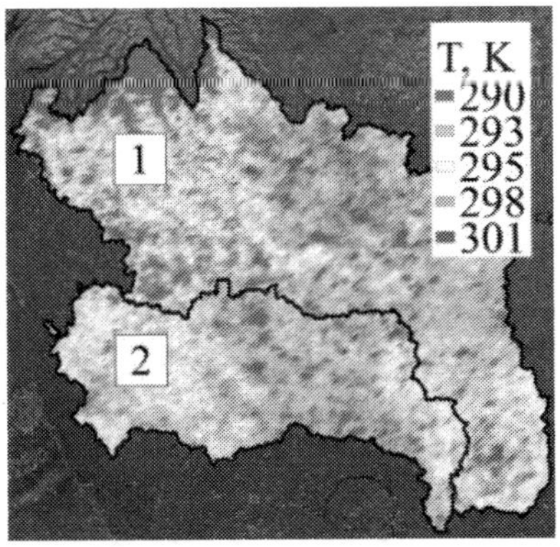


**Fig. 2.** Surface temperature distribution in the study area in July 2018. 1 - basin of the Nizhnyaya Tunguska river, 2 - basin of the Podkamennaya Tunguska river

For a detailed analysis of the dynamics of surface temperature, a polygonal GIS layer of smaller taxa was used, consisting of 18 river basins on a scale of 5 nesting levels out of 12 according to the HydroSHEDS classification (Fig. 3).

### 2.2 Gravimetric data

To assess the effect of thermal anomalies on the water cut of the study area, an archive of gravimetric data was formed based on information received from the GRACE and GRACE-FO satellite systems. This data was obtained from the site https://podaac-tools.jpl.nasa.gov/. They are provided with aggregation by month, contain information about the height of the water layer above the geoid and are expressed as the EWH (Equivalent Water Height) parameter. The data has a spatial resolution of 1x1 degree. The data archive contains information from January 2013 to December 2019.

## 3 Results and discussion

Figure 3 shows the results of classification, which reflect the degree of disturbance (γ) of smaller river basin within the basins of the Nizhnyaya Tunguska river and Podkamennaya Tunguska river for 2001, 2009, and 2019.

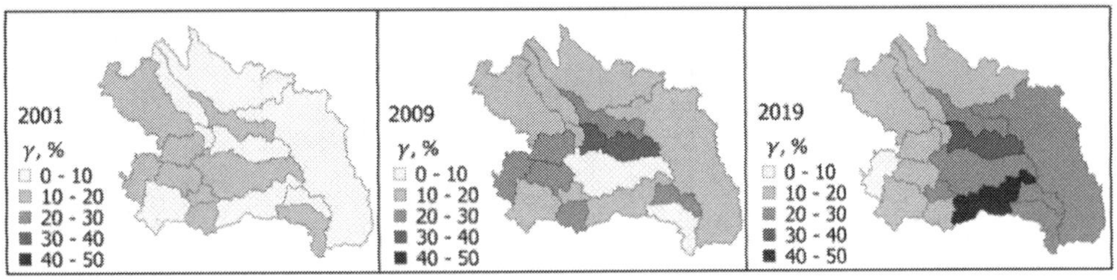


**Fig. 3.** The degree of disturbance (γ) of the study area in 2001, 2009 and 2019

Based on the analysis of the values of the underlying surface temperature, it was found that from 2001 to 2009 the total area of disturbances increased by 7.77% or 36752.1 km$^2$ within the basin of the Nizhnyaya Tunguska river. At the same time the total level of disturbances within the basin of the Podkamennaya Tunguska river increased by 4.7% or 11,280 km$^2$. During the next period of 2009–2019 this characteristics increased by 10.34% or 48908.2 km$^2$ and by 17.19%, or 41,256 km$^2$, respectively. The total percentage of areas characterized with signs of disturbed vegetation cover was 9.36% or 66,736.8 km$^2$ (in 2001), 16.09% or 114,721.7 km$^2$ (in 2009) and 28.8% or 205344 km$^2$ (in 2019) (Fig. 4).

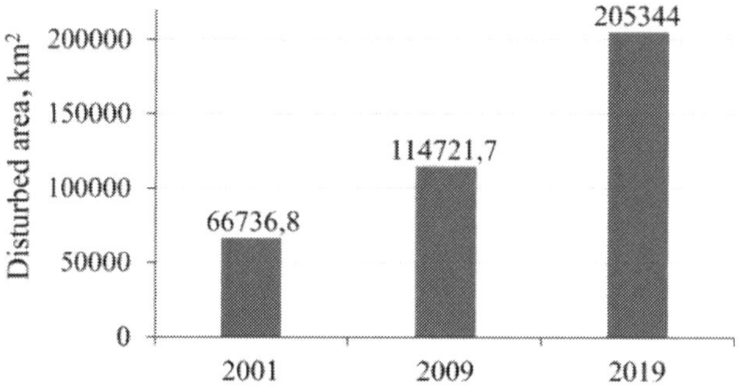


**Fig. 4.** Changes in the areas of disturbance of the underlying surface of the study area during two 10-year periods.

Wildfires make the greatest contribution to the appearance of disturbance of the underlying surface. This applies to all 18 surveyed polygons (which are the basins of smaller rivers, see fig. 3). Also, an anthropogenic factor cannot be excluded, namely logging activities, mining, etc.

In order to determine whether the degree of disturbance of the forest cover has a relationship with the water cut in the study area, the variance values for the temperature of the underlying surface and the EWH parameter for the study area were calculated. The overall average surface temperature and EWH parameter were also calculated for the territory of the two studied river basins.

**Table.** Statistical parameters of EWH and temperature data for the territory of two basins with aggregation by season

| Season | Year | GRACE EWH, cm | | MODIS t, K | |
|---|---|---|---|---|---|
| | | σ | M | σ | M |
| Winter | 2013 | 0,017 | -0,026 | 3,0 | 242,1 |
| Spring | | 0,031 | 0,025 | 3,2 | 274,7 |
| Summer | | 0,022 | -0,065 | 2,3 | 294,6 |
| Autumn | | 0,011 | -0,041 | 3,5 | 257,4 |
| Winter | 2014 | 0,017 | 0,051 | 4,8 | 241,3 |
| Spring | | 0,033 | 0,106 | 4,0 | 269,3 |
| Summer | | 0,029 | -0,008 | 1,9 | 293,1 |
| Autumn | | 0,021 | -0,037 | 4,4 | 262,2 |
| Winter | 2015 | 0,032 | 0,067 | 5,1 | 245,6 |
| Spring | | 0,031 | 0,125 | 3,0 | 268,9 |
| Summer | | 0,029 | -0,067 | 1,9 | 291,7 |
| Autumn | | 0,019 | -0,054 | 2,3 | 282,2 |

| | | | | | |
|---|---|---|---|---|---|
| Winter | 2016 | 0,027 | 0,021 | 3,0 | 247,6 |
| Spring | | 0,025 | 0,055 | 3,5 | 267,3 |
| Summer | | 0,034 | -0,071 | 1,8 | 294,2 |
| Autumn | | 0,015 | -0,089 | 2,7 | 248,0 |
| Winter | 2017 | 0,032 | -0,044 | 4,1 | 240,9 |
| Spring | | 0,028 | 0,035 | 4,2 | 269,5 |
| Summer | | 0,016 | 0,027 | 3,4 | 294,8 |
| Summer | 2018 | 0,033 | -0,051 | 2,4 | 296,5 |
| Winter | 2019 | 0,022 | 0,003 | 2,9 | 242,9 |
| Spring | | 0,025 | 0,037 | 4,0 | 268,7 |
| Summer | | 0,026 | -0,036 | 2,2 | 294,8 |
| Autumn | | 0,020 | -0,072 | 3,5 | 264,4 |

It can be seen from the table that some time intervals are missing, for example, autumn 2017 and the whole of 2018, with the exception of the summer period. This is due to the lack of data from the GRACE system due to its unstable operation.

The table shows that the maximum values of the EWH parameter correspond to the spring period, the minimum values of this parameter are observed in the autumn period.
In order to find out that an increase in the temperature of the underlying surface affects the state of the permafrost layer, the level of groundwater and the dynamics of river flows, it is necessary to compare the series of data on the temperature of the underlying surface with the series of data on the EWH parameter.

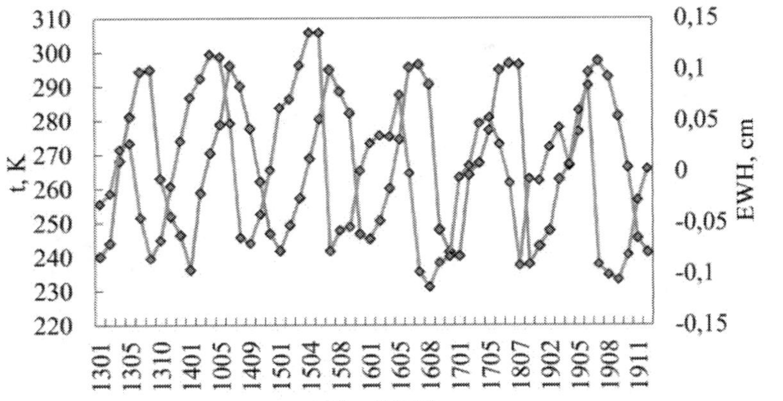


**Fig. 5.** Annual temperature fluctuations (t) and EWH

Figure 5 shows the monthly average values of the EWH parameter and surface temperature for the territory of two river basins in the period from January 2013 to December 2019. These surface temperatures have a positive trend, which is consistent with the results presented above. The negative trend in the values of the EWH parameter for the period under study indicates a decrease in the total water cut of the territories of the two river basins. In combination with a positive temperature trend, it can be assumed that there is a relationship between an increase in the temperature of the underlying surface due to an increase in the number and areas of disturbed forest areas and a negative EWH trend, which indicates a decrease in groundwater reserves in the study area. To check the relationship between the runoff data from the Podkamennaya Tunguska river and the GRACE and GRACE-FO gravimetric measurements for the territory of the corresponding river basin, we compared the values of the EWH parameter over the study area and the river runoff values. The graph of the correlation field (Fig. 6) demonstrates the general relationships between the parameters of "river runoff" – "EWH parameter".

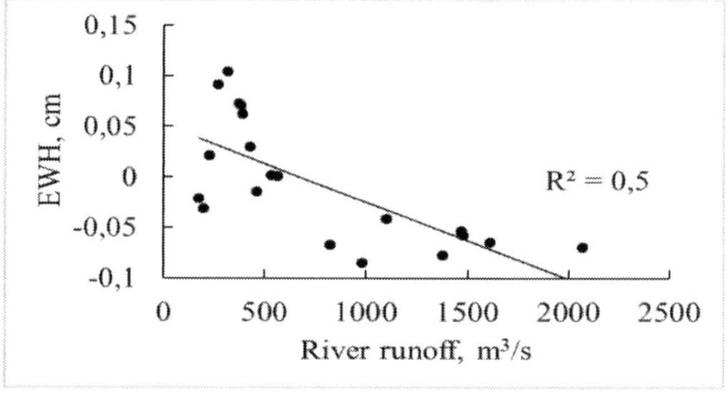


**Fig. 6.** Correlation field of the river runoff rate of the Podkamennaya Tunguska river and the EWH parameter

Figure 6 shows that the peak values of the river runoff correlate with the maximum value of the EWH parameter. The in phase of the minima of both investigated parameters was also confirmed.

## 4. Conclusions

The remote sensing data in the IR range makes it possible to assess the degree of disturbance of territories by analyzing the surface temperature and comparing the detected temperature anomalies at the sites of damaged vegetation with the background values of the underlying surface temperature. As studies show, thermal anomalies are significant even after 10 – 15 years [2, 6] of recovery processes, which makes monitoring the state of forest cover using satellite data in the infrared range promising.

According to the results of calculations in the study area, there is an increasing trend in the areas of disturbances of the underlying surface over the past 20 years. In 2001, the total area of disturbance was 66736.8 $km^2$, and it was rising up to 114,721.7 $km^2$ in 2009 and up to 205344 $km^2$ in 2019.

Wildfires make the greatest contribution to the appearance of disturbance of forests and on–ground cover of the region. A technogenic factor, for example, logging activities, mining, etc., also directly affects the distribution of thermal anomalies, which are confidently detected by using of remote sensing data in the thermal IR range.

The combined use of gravimetric data and data on the temperature of the underlying surface allows, in a first approximation, to substantiate the applicability of gravimetric survey data for analyzing the dynamics of river runoff changing under conditions of large-scale post–fire damage to the larch forests in the permafrost zone.

## Acknowledgments

This research was funded by the Russian Foundation for Basic Research (RFBR) and Government of the Krasnoyarsk krai, and Krasnoyarsk krai Foundation for Research and

Development Support, grant number 18-41-242003 “Modeling and satellite monitoring of effects from thermal anomalies of the underlying surface in the seasonally thawed soil layer of the permafrost zone of Siberia”, and grant number 20-44-242002 “Instrumental monitoring of physical properties and numerical modeling of the state of technogenically disturbed soils in Siberia”, and Grant of the East Siberian Oil and Gas Company aimed at supporting scientific research of applied importance in 2020.

## References

1. E.A. Loupian, S.A. Bartalev, I.V. Balashov, V.A. Egorov, D.V. Ershov, D.A. Kobets, K.S. Senko, F.V. Stytsenko, I.G. Sychugov, Sovr. Probl. DZZ Kosm 6, 158 (2017)
2. E.I. Ponomarev, T.V. Ponomareva, Cont. prob. of ecol., 4, 420 (2018)
3. K.V. Krasnoshchekov, A.V. Dergunov, E.I. Ponomarev, Sovr. Probl. DZZ Kosm 2, 87 (2019)
4. N.D. Yakimov, E.I. Ponomarev, E3S Web of Conferences **149**, 03008 (2020)
5. A.A. Knorre, AV. Kirdyanov, A.S. Prokushkin, P.J. Krusic, U. Büntgence, Sci. Total Environ **652**, 314 (2019)
6. E.I. Ponomarev, O.V. Masyagina, K.Y. Litvintsev, T.V. Ponomareva, E.G. Shvetsov, K.A. Finnikov, Forests **11**, 8, 790 (2020)
7. E.I. Ponomarev, T.V. Ponomareva, A.S. Prokushkin, Water. **11**, 6, 1146 (2019)
8. M. Helbig, J.M. Waddington, P. Alekseychiket [et al.], Nat. Clim. Change 10 555 (2020)
9. S. T. Im, V. I.Kharuk, Izvest., Atm. and Oc. Phys. **51**, 8 (2015)
10. B. Lehner, G. Grill, Hydrol. Process **27** 15. 2171 (2013)
11. Z. Wan, S. Hook, G. Hulley, MODIS land surface temperature products users’ guide, ICESS (2015).

# The use of satellite data for monitoring rivers in the Amu Darya basin

*Ildar* Mukhamedjanov*, *Anna* Konstantinova and *Evgeniy* Loupian

Space Research Institute of the RAS (IKI), 84/32 Profsoyuznaya, Moscow, Russia, 117997

**Abstract.** The paper explores the organization of satellite monitoring of water bodies in Central Asia on the example of the Amu Darya river in a specialized satellite monitoring system EcoSatMS (http://suvo.geosmis.ru). The potential and prospects of using the new technology of virtual gauging stations to control the state of water bodies, as well as restoring the values of daily runoff at a point remote from the ground station for some distance are highlighted. Evaluation of the effectiveness of the implementation of the virtual gauging stations technique in flow of automated calculations allows to take the research as a basis for further development of individual tools for analyzing satellite observation time series and the EcoSatMS in general.

## 1 Introduction

The typical feature of the water resources of the Central Asian region are transboundary rivers. The Amu Darya is the most affluent river in Central Asia, which is at the same time the basis of the water potential for the southern part of the region. The transboundary waters of the Amu Darya play an important role in irrigation and agriculture, in particular for Uzbekistan. Under such conditions, satellite data help to objectively and timely assess the state of the certain parts of the river bed and water bodies of other origin (lakes and water reservoirs). Modern remote sensing systems make it possible to organize the acquisition of observation series of natural and anthropogenic objects with a time frequency of several days. To solve the problems of complex satellite monitoring of water bodies in Uzbekistan and Central Asia in general, a special satellite monitoring system EcoSatMS (Ecosystem Satellite Monitoring service, http://suvo.geosmis.ru) has been developed at the Space Research Institute of the Russian Academy of Sciences [2].

The system is based on the "IKI-Monitoring" center for collective use infrastructure [1]. is the part of the Sozvezdie-Vega services family (http://sozvezdie-vega.ru), it is developed and maintained at the Space Research Institute of the Russian Academy of Sciences (IKI). The map interface of the service is adapted for the territory of Uzbekistan and neighboring countries, the basic cartographic layers of the region boundaries, road networks, water bodies, settlements, the boundaries of the catchments of rivers in Central Asia are integrated [3]. The long-term data archives with a depth of more than 30 years

* Corresponding author: ildarmsu@gmail.com

make it possible to form satellite observation time series, as well as to organize online access to permanently updated catalogs of satellite data and calculated products.

# 2 Materials and methods

## 2.1 Data processing

A specialized tool for remote monitoring objects has been introduced into the water resources analysis system, which implements the technology of "virtual gauging stations". The virtual gauging station is a polygon of any shape, which is saved to the database by user in an interactive mode (Fig. 1). At the same time, it is possible not only to select the period for calculating the parameter of the virtual gauging station based on the satellite images from the long-term archive, but also to analyse these images in automatic mode as the satellite data appears.

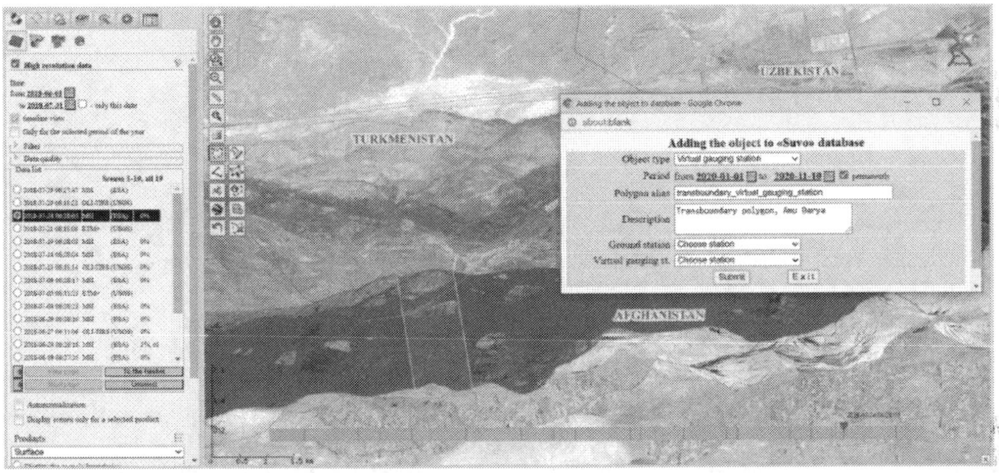


**Fig. 1.** Creating a virtual gauging station in EcoSatMS

Statistics collection for the space gauging station in automatic mode takes several minutes. The goal of observations calculation is the area of the water surface, computed by the modified difference water index, or MNDWI (modified normalized water index) [4]. The satellite optical images of systems OLI, TIRS (Landsat-8), MSI (Sentinel-2A, -2B), TM (Landsat -4, -5) and ETM + (Landsat-7) are used. Pixels with MNDWI value greater than 0 are taken as the water surface. The obvious advantages of this approach when assessing the state of water bodies are low resource consumption, independence from geographic location, efficiency and scalability. Observation time series obtained over a certain period are further processed using the collaborative time series analysis tool.

## 2.2 The application of the virtual gauging stations technique

The process of ground measuring stations settlement for assessing the river features often causes a number of difficulties due to the location complex terrain morphology or the features of political boundaries. Unlike an expensive physical object, setting up a polygon in an online service does not require material costs and is reduced to the work of an operator. One of the key figures in assessing the hydrological situation on the river is daily runoff. To use the data of the virtual gauging station to estimate the water discharge per day, it is necessary to identify the relationship between the ground figure and the calculated water surface area. Within the framework of the current research, a number of experiments were carried out, which made it possible to find a linear relationship of the area of water surface area on daily runoff at the Kerki gauging station in Turkmenistan. The results of data analysis at Kerki for 2018 and at the virtual gauging station 1 km away from the station are established in Fig. 2. Performed calculations were carried out using the online tool for time series analysis in EcoSatMS. At this step, the features below were also calculated:

$R = 0.957$ – Pearson coefficient of correlation

$R^2 = 0.916$ – determination coefficient

$MSE = 0.006$ – mean squared error value

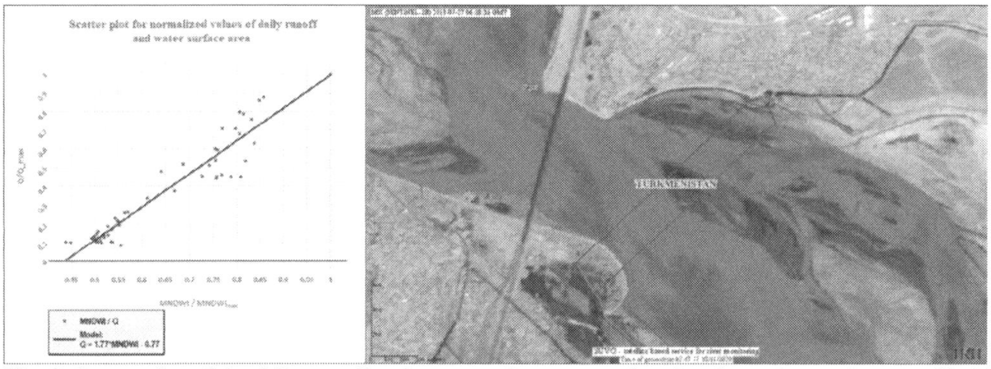


**Fig. 2.** Scatter plot of the daily runoff and water surface area at the virtual gauging station near Kerki and the location of the polygon

Time series of daily runoff normalized to a maximum runoff in 2018 and the figures of the water surface area normalized to a maximum of area in 2018 were used as a dataset.

After the linear relationship was found at the virtual gauging station closest to Kerki, the question of how random such a relationship can be from the point of view of the polygon positioning was investigated. In addition, it is necessary to establish whether it is possible by sequential linking of virtual gauging stations with each other and using a linear relationship to preserve the correlation, moving away from the ground station. For this, when a new polygon is created, EcoSatMS provides a feature to link the current virtual gauging station either with another one or with the ground station. This sequence of related polygons is called "virtual gauging stations net". As there is a correlation between the daily runoff and the water surface area figures, a linear model was used to calibrate the virtual gauging stations.

## 2.3 Virtual gauging stations net calibration

Let Q denote the daily runoff and S denotes water surface area figures. To produce a net according to the following technique it is necessary to save at each iteration the coefficients a and b of the linear equation:

$$Q = a * S + b$$

Calibration of the virtual gauging station means preserving the linear model coefficients and the id of the previous polygon for further daily runoff time series restoring.

The following steps are performed for each object:

1st iteration. The closest polygon to Kerki is linked to to ground station:

$$Q_0 \sim Q_1 = a_1 * S_1 + b_1$$

n+1[st] iteration. A link between the current polygon and the previous one is created. The following calculations are performed:

1. $S_n = a_{n+1} * S_{n+1} + b_{n+1}$
2. $Q_n = a_n * S_n + b_n * Q_n \sim Q_{n+1} = (a_n * a_{n+1}) * S_{n+1} + a_n * b_{n+1} + b_n$
3. Daily runoff figures restoring using linear regression:

$$Q_{n+1} \sim c_{n+1} * Q_0 + d_{n+1}$$

At each step of a virtual gauging stations network construction the next figures are calculated and analyzed:

• S, ha – polygon area;

• $R$ ($Q_{true}$, $Q_{restored}$) – correlation coefficient of the series of true and restored daily runoff values;

• $R$ ($Q_{true}$, $S_{MNDWI}$) – correlation coefficient of the series of the water surface area with true daily runoffs figures (criterion to select the best polygon at the 1[st] iteration);

• $F$ ($Q_{true}$, $Q_{restored}$) – ratio of variances of true and restored daily runoff series (Fisher's test). F-value is considered as a key value to find appropriate polygon at n+1[st] iteration ($n > 1$);

• $MSE$ ($Q_{true}$, $S_{MNDWI}$) – mean squared error for normalized daily runoff and normlized water surface area figures at the current virtual gauging station;

At each iteration, the diagram shown in Fig. 3, where the polygon set 6 km from the ground station and calibrated by the nearest to Kerki polygon are considered.

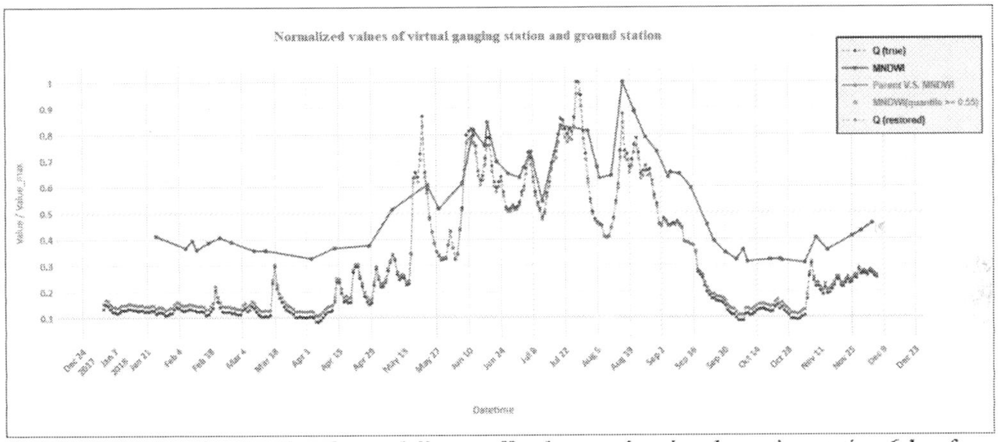


**Fig. 3.** The chart of restored and true daily runoff values at the virtual gauging station 6 km from Kerki (the 2[nd] iteration of the network construction: Kerki → 1 km → 6 km)

# 3 Results

## 3.1 Polygons settlement rules

A detailed analysis of shape effect, area, orientation of polygons and terrain morphology allowed to construct a quality network of 12 virtual gauging stations stations, which allows to save the correlation of the true measurements with the restored ones above 0.85 at a distance of over 25 km from the ground station. In addition, it was also found that the maximum distance at which the calibration proposed in the paper can preserve the correlation between the restored and true of daily runoff values depends on the variability of the river course structure and irrigation system throughout the entire river section. In this case, at a distance of 22 km upstream from Kerki a channel way takes its origins from Amu

Darya to north. As a result, the daily runoff values will differ quantitatively from those measured at the ground station.

As a result of analyzing the shape and calculating statistics for more than 60 polygons, the following rules were formed to select the virtual gauging stations to combine them further into a network:

1. It is necessary to avoid getting into the polygon zones formed by long-term shore erosion between river streams during the low-water period;
2. The polygon should not be exposed to flooded fields at the boundary area;
3. Areas with stable snow cover should be excluded from the polygon (snow and river ice are recognized as water surface when calculating the water surface area by MNDWI) as they falsely increase the area;
4. Pittings in river bed or river terrace containing standing water make noise when calculating the water surface area;
5. The direction of the secant vector for a part of the polygon should be perpendicular to the vectors of partial flows of the river section [5].

### 3.2 Test results for 2014 and 2016

Time series of daily runoff for 2014 and 2016 were selected to check the quality of the virtual gauging stations network constructed using data of 2018. Fig. 4 performs a chart of the Fisher's test result for virtual gauging stations at different distances from Kerki. Empirically found $F = 2.5$ was taken as a threshold.

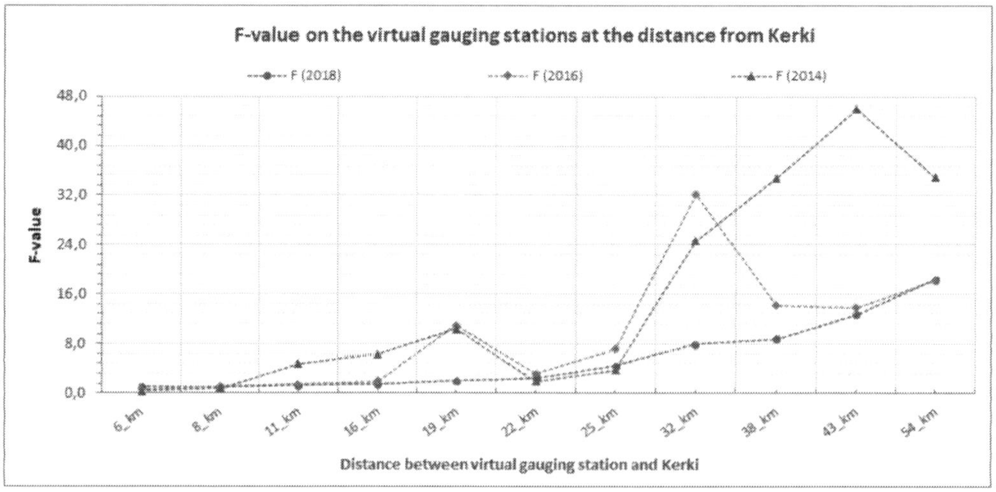


**Fig. 4.** F-value chart for virtual gauging stations at different distances from Kerki

The chart shows that the result model cannot be used in 2016 at a distance greater than 16 km, and in 2014 further than 8 km. Such indicators may perform the difference in the water regime and water amount in different years. It means that the model should be more stable in the interannual aspect and take into account more factors.

## 4 Conclusion

The described technique at the current stage helps to form a virtual gauging stations net and restore daily runoff time series at different points of the Amu Darya. Expanding the package of test data and data network calibration will help improve the accuracy of finding linear regression coefficients, which can make the current model more stable. To improve the accuracy and stability of the method in the interannual aspect, it is necessary to test more complex nonlinear models, as well as to include more features to model. The results obtained are the basis for implementing virtual gauging stations nets as an entity in EcoSatMS and automating the process of such networks construction.

The research was carried out using the capabilities of the «IKI-Monitoring» center for collective use [1] with the support of the project No. 19-37-90114 of the Russian Foundation for Basic Research.

## References

1. Loupian E.A., Proshin A.A., Bourtsev M.A., Kashnitskiy A.V., Balashov I.V., Bartalev S.A., Konstantinova A.M., Kobets D.A., Mazurov A.A., Marchenkov V.V., Matveev A.M., Radchenko M.V., Sychugov I.G., Tolpin V.A., Uvarov I.A. Experience of development and operation of the IKI-Monitoring center for collective use of systems for archiving, processing and analyzing satellite data // Sovremennye problemy distantsionnogo zondirovaniya Zemli iz kosmosa. Vol. 16. No. 3. P. 151-170. (2019) DOI: 10.21046/2070-7401-2019-16-3-151-170.
2. Mukhamedjanov I.D., Konstantinova A.M., Uvarov I.A. Capabilities of the "IKI-Monitoring" Center for collective use in organizing the satellite monitoring of the Central Asian Region // Proceedings of the International conference InterCarto/InterGIS, GI support of sustainable development of territories. Vol. 26. No. 2. P. 376–383. (2020). DOI: 10.35595/2414-9179-2020-2-26-376-383
3. Mukhamedjanov I.D., Konstantinova A.M., Loupian E.A., Gafurov A.A. Approaches to the construction of the satellite service system for Amu Darya river and its tributaries // Proceedings of the XVII All-Russian Open Conference «Actual Problems of Remote Sensing of the Earth from Space». P. 111. (2019). 11-15 November 2019, Moscow, Russia. DOI: 10.21046/17DZZconf-2019a
4. Pogorelov A.V., Lipilin D.A., Kurnosova A.S. Satellite monitoring of the Krasnodar reservoir // Geographical bulletin. No. 1(40). P. 130–137. (2017). DOI: 10.17072/2079- 7877-2017-1-130-137
5. Bykov V.D., Vasilyev A.V. Hydrometry. Leningrad, Gidrometeoizdat. (1977). 448 P.

E3S Web of Conferences **223**, 03009 (2020) https://doi.org/10.1051/e3sconf/202022303009
*RPERS 2020*

# About the features of the time course of the average annual air temperature in the territory of the Debed river basin (Armenia)

*Varduhi* Margaryan[1,*], *Gennady* Tsibulskii[2], *Ksenia* Raevich[2]
[1] Yerevan State University, Department of Physical Geography and Hydrometeorology, Faculty of Geography and Geology, 0025, Alek Manoukian 1 st., Yerevan, Armenia
[2] Siberian Federal University, Institute of Space and Information Technologies, Chair of Systems of Artificial Intelligence, 660074 Kirenskogo St. 26, Krasnoyarsk, Russia

**Abstract.** The article examines the features of the time course of the average annual air temperature in the Debed river basin in Armenia. As a starting material, we used daily data of actual observations of the temperature of the surface air layer for a year in the Debed river basin. The study was carried out at 6 meteorological stations in the Debed river basin based on long-term observation data series from 1930 to the present (2018). Analysis of the trend lines of temporal changes in air temperatures shows that at all meteorological stations currently operating on the territory of the basin, there is mainly a tendency for an increase in temperatures of annual values.

## 1 Introduction

Air temperature is characteristics of situation on of links of climatic system (atmosphere). The air temperature has a main role in the forming of atmosphere circulation, evaporation and moisture regime of territory. Clarifying and estimation of regularities of temporal distribution of air temperature has importance, especially for more accurate definition of thermal balance, for productive using of thermal resources. That is why the air temperature study is an important task and caught our attention.

The purpose of this study is to solve the following tasks: to identify, analyze and evaluate the features of the territorial distribution of the mean annual air temperature, to estimate their time course over the past 80-90 years in different conditions of the Debed river basin.

## 2 Study areas, Data and Methods

To solve the set tasks, the corresponding studies and published works [3, 8-9] served as a theoretical basis in the work. As a starting material, the work used the daily factual data of the «Center for Hydrometeorology and monitoring» of the Republic of Armenia for a period of 80 years and more (1930–2018) 6 meteorological stations.

* Corresponding author: vmargaryan@ysu.am

The Debed river basin is located in the north of the Republic of Armenia, occupying an area within the republic of 3790 sq.km (4080 km2 – outside of it). Here, in the lowest point (375m) of the territory of the Republic of Armenia. The basin is distinguished by a rugged relief, there are canyons reaching 300m depth (the Debed river canyon), and separate massifs reaching more than 2500m in the height. The difference in altitude exceeds 2800m. In the basin, one can trace the mountain-forest, mountain-steppe and alpine climatic zones, only in the deep gorges of the extreme northeastern part of the basin a moderately warm climate of dry steppes is formed.

The following methods were used in the work: mathematical-statistical, correlation, extrapolation, interpolation, analysis, synthesis, analogy.

## 3 Results and discussion

The study area is characterized by an uneven distribution of air temperature (table 1), which is mainly due to the geographical latitude of the area, general and local circulation of the atmosphere, the nature of the active surface, radiation energy and orographic features. On the territory of the Debed river basin, from an altitude of 451 to 2066 m, according to actual observations, the average annual temperature is positive. Long-term average annual values of air temperature for the period from the 30th of the XX century to 2018 on average ranges from 3,74 °C (Pushkin pass) to 12,3 °C (Bagratashen).

**Table 1.** Values of average annual air temperature at meteorological stations for the entire period of their operation

| Meteorological stations | Bagratashen | Tashir | Odzun | Stepanavan | Pushkin pass | Vanadzor |
|---|---|---|---|---|---|---|
| Height, m | 451 | 1507 | 1105 | 1397 | 2066 | 1376 |
| Air temperature, °C | 12,3 | 6,14 | 9,34 | 7,21 | 3,74 | 8,27 |

As a rule, with the absolute height of the area, the annual air temperature naturally decreases (table 2). It turned out that in the study area with an altitude of every 100m, the annual air temperature decreases on average by 0,54 °C. On the basis of this connection, it is possible to study the regime of air temperature and precipitation in unexplored and poorly studied areas of the discussed territory, as well as to build a map of their territorial distribution.

**Table 2.** Values of the average annual air temperature in altitude zonation

| Altitudinal zones, m | 600 | 800 | 1000 | 1200 | 1400 | 1600 | 1800 | 2000 | 2200 | 2400 |
|---|---|---|---|---|---|---|---|---|---|---|
| Air temperature, °C | 11,7 | 10,7 | 9,6 | 8,5 | 7,4 | 6,3 | 5,3 | 4,2 | 3,1 | 2,0 |

Changes in the average annual air temperature were also estimated for the entire territory of the Debed basin (tig. a, fig. 1), and for each of the currently operating meteorological stations separately (tig. b,c,d fig. 4). Since the meteorological stations began their activities at different times (for example, the Pushkin pass meteorological station has been operating since 1963), the average air temperature for the year and the annual amount of precipitation for the entire territory have been calculated since 1964. In the Debed river basin as a whole, a stable positive trend of the average annual air temperature was revealed. For the period from 1964–2018 the rate of annual warming will be +0.296 °C/10 years. A significant increase in annual temperatures is observed especially over the period 1993–2018. On average, the air temperature in the Debed basin from 1964 to 2018 increased by 1,65 °C over the year.

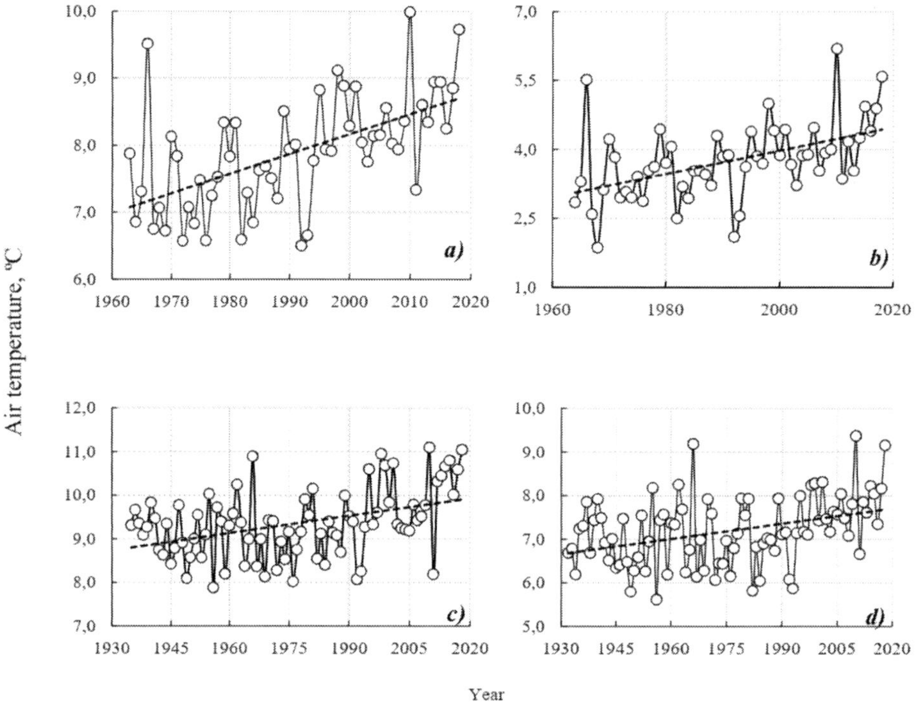


**Fig. 1.** Values of the average air temperature for the year in the Debed river basin as a whole (a) and for meteorological station at different heights : Pushkin pass – 2066m (b), Odzun – 1105m (c), Stepanavan – 1397m (d).

In the Debed river basin as a whole, over the past 60 years, 1972, 1976, 1982, 1992 are distinguished by relatively low annual temperatures (less than 6,5–6,6 °C). Since 1993, there has been a sharp rise in temperature. Comparatively high annual temperatures (more than 9,5 °C) stand out in 1966, 2010 and 2018. The warmest years were also in 1966 and 2010 on the territory of the high mountain landscapes of the northeastern Caucasus [1].

This pattern is typical both for other regions [11–13), and for the whole territory of Armenia [3, 10], other regions of our planet, including the South Caucasus [6], Eastern Georgia [2], the territory of high-mountainous landscapes of the north-eastern Caucasus [1], the Russian Federation [7], the territory of all CIS countries [5], the Ulyanovsk region [14], the landscape of the North Caucasus [4]. According to Armenia's fourth national communication on climate change [2], in hate Republic of Armenia over the period of 1929-1996, the annual mean temperature increased by 0,4 °C, during 1929–2007 – by 0,85 °C, during 1929-2012 – by 1,03 °C, and during 1929–2016 – by 1,23 °C. According to the METRAS model and RCP8.5 scenario in the territory of Armenia expected the annual mean temperature increased by 1,6 °C during 2011–2040, by 3,3 °C during 2041–2070, by 4,7 °C during 2071–2100.

Table 3. The projected values of average annual air temperature (°C) in the territory of Armenia by different altitudinal zones (m), according to the METRAS model and RCP8.5 scenario [3]

| Altitudinal Zones, m | 1961–1990 | 2011–2040 | 2041–2070 | 2071–2100 |
|---|---|---|---|---|
| < 800 | 11,2 | 12,8 | 14,5 | 15,9 |
| 800–1000 | 10,8 | 12,4 | 14,1 | 15,5 |
| 1000–1500 | 8,4 | 10,0 | 11,7 | 13,1 |
| 1500–2000 | 5,5 | 7,1 | 8,8 | 10,2 |
| 2000–2500 | 3,3 | 4,9 | 6,6 | 8,0 |
| 2500–3000 | 1,6 | 3,2 | 4,9 | 6,3 |
| > 3000 | -0,7 | 0,9 | 2,6 | 4,0 |
| Armenia | 5,5 | 7,1 | 8,8 | 10,2 |

The change of air temperature will its inevitable consequence on change of components of hydrothermal balance study area, on a violation of the ecological balance of natural ecosystems, as well as the social, environmental and economic development of study area. Therefore, to adapt to changing temperature needs an ecosystem approach, for mitigation – implementation of complex measures for adaptation.

## 4. Conclusions

As a result of the research came to the following conclusions:

– The distribution of air temperature in the studying area is uneven: the average annual air temperature ranges from 3,74 to 12,3 °C. The air temperature decreases with the terrain height: the vertical gradient is 0,54 °C/100m.

– Long-term fluctuations in the average air temperature for the year are everywhere characterized by positive trends. On average, in the Debed river basin from 1964 to 2018, the air temperature increased by 1,65 °C over the year. The warmest years were observed in 1966, 2010 and 2018.

– It is necessary to develop strategy programs for adapting the consequences of changes in the average annual air temperature, which will serve as a stimulus for the long-term development of the region's economy.

– The results obtained can be used in planning the development of the economy on the territory of the Debed river basin, in providing climatic services to the branches of the national economy, in adjusting building standards.

## References

1. A.A. Abdulzhalimov, Z.V. Ataev, V.V. Bratkov Izvestiya Dagestanskogo gosudarstvennogo pedagogicheskogo universiteta. Journal Dagestan State Pedagogical University. Natural and Exact Sciences. 2, 9 (2015)
2. A. Amiranashvili, T. Matcharashvili, T. Chelidze Journal of Georgian Geophysical Society, Issue (A), Physics of Solid Earth. 15, 21 (2011-2012)
3. Armenia's Fourth National Communication on Climate Change. (Yerevan, UNDP Armenia. 2020)
4. Z.V. Atayev, V.V. Bratkov The South of Russia: ecology, development. 1, 17 (2014)
5. M.Yu. Bardin, T.V. Platova, O.F. Samokhina Proceedings of the FSBI «Hydrometeorological Center of Russia». 358, 23 (2015)
6. Building Resilience to Climate Change in South Caucasus Agriculture / Nicolas Ahouissoussi, James E. Neumann, and Jitendra P. Srivastava, Editors. (International Bank for Reconstruction and Development, 2014)
7. G.V. Gruza, E.Ya. Rankova Observed and expected climate changes over Russia: surface air temperature. (Obninsk: «VNIIGMI-MTSD», 2012)

8. IPCC: Climate Change 2013: The Physical Science Basis. Cambridge University Press, Cambridge, United Kingdom and New York, NY, USA (2013)
9. K.S. Kirillina, V.A. Lobanov Proceedings of the Russian State Hydrometeorological University (RSHU). 38, 15 (2015)
10. V.G. Margaryan The conference Global Climate Observation: the Road to the Future. (Amsterdam, 2-4 March 2016)
11. V.G. Margaryan Moscow University Bulletin. Series 5. Geography. 2, 5 (2019)
12. V.G. Margaryan Visnyk of V.N. Karazin Kharkiv National University, series «Geology. Geography. Ecology». 50, 11 (2019)
13. V.G. Margaryan, N.I. Samvelyan Vestnik Voronezskogo gosudarstvennogo universiteta. Seria Geografia. Geoekologia. 4, 8 (2019)
14. Yu.P. Perevedentsev, R.B. Sharipova Bulletin of the Udmurt University. Series «Biology. Earth Sciences». 1, 9 (2012)

# Determination of the earth's surface deformations in the "Kiyzassky" open pit according to Sentinel-1 Data

*Leonid* Mikov*, *Semion* Popov

Federal Research Center for Information and Computational Technologies, Geoinformation Modelling Lab, 630090 Novosibirsk, Russia

**Abstract.** The paper deals with the issues of assessment of the condition and changes in the land surface on the territory of the Kiizassky open pit (Kemerovo region) because of the landslide that occurred in June 2019. The application of the multi-pass series of Sentinel-1 satellite radar data using the Small Baseline Subset (SBaS) method to determine the Earth surface displacement dynamics using constructed vertical displacement maps is demonstrated..

## 1 Introduction

Currently Kuzbass is one of the main mining regions in Russia. High technogenic load associated with the annual increase in coal production leads to the redistribution of stress-strain state of significant volumes of rock mass with adverse consequences for the environment [1]. Deformations of the Earth's surface in such regions, resulting from uneven subsidence and horizontal movements of rocks, reach large territories and are the source of emergency situations, which, in turn, may harm not only the environment and the operating mode of the enterprise, but also human lives. For example, in June 2019, a landslide occurred at one of the Kiyzassky open-pit mine. As a result, the supports of nine power transmission lines were damaged and the bed of the Bolshoi Kiyzas River was blocked by 750 meters in length and about five meters in height [2] (Fig. 1) [3].

## 2 InSAR data processing methods

Today, satellite radar data are actively used to determine the deformations of the earth's surface. Their advantage is the relatively low price per square kilometer, as well as the independence of the shooting from weather conditions and time of day. To process such data, 3 main methods have been developed:

- differential interferometry (DinSAR). Used to process two images

* Corresponding author: mikov@ict.sbras.ru

• method of persistent scatterer (PS). It is used for point estimation of deformations based on a multi-pass series of images

**Fig. 1.** Photo of the Kiyzassky open pit stockpile before the landslide (left) and after the landslide (right)

• small baseline subset method (SBaS). Used for areal strain assessment based on a multi-pass series of images.

The principle of radar interferometry is as follows. Each pixel of the satellite image contains information about the real and imaginary part of the wave reflected from the part of the Earth's surface and received by the satellite. Thus, the satellite image has two components: real and imaginary part, which can be converted into the amplitude (brightness) and phase components of the image. Note that a typical size of the Earth surface area in one pixel of the image exceeds a square meter, which makes it impossible to determine millimeter offsets by the brightness component of the image. For this purpose, the analysis of phase components of the image set is used.

Based on a set of phase components of SAR images received by the satellite at different points in time (with periodicity, for example, one image per day), the change in phase component is calculated over time. The phase component change is due to a linear dependence with the difference in the course of the wave reflected from the surface, and thus it is possible to measure the vertical displacements of objects on the Earth's surface, comparable to the signal length of the satellite, which has the order of a few centimeters. [4]

The main difference between the methods is the number of source data to be processed. For example, for the method of differential interferometry a pair of radar images taken with a small time interval (from several days to a couple of weeks) is used. As a result, we get a map of the Earth's surface displacements that occurred between shots. The SBaS method requires at least three images, while the PS method requires at least thirty. The last two methods allow us to estimate the dynamics of the Earth surface changes over a long period of time (from several months to several years). Let us consider all these methods in detail.

Differential interferometry is a classical pairwise method for calculating area displacements. It includes the following stages: interferogram formation, filtering and coherence calculation, phase unwrapping, orbital correction and refinement, phase to displacement recalculation. At the output, we get a map of the Earth's surface displacements.

PS (Persistent Scatterers) method - calculation of displacements of point targets which are permanent reflectors for radar satellite. It allows to measure detailed displacements on infrastructure objects. Used, as a rule, for areas with urban terrain. Includes the same stages as the SBaS method. At the output we also get a map of the average velocities of changes in surface displacements.

SBaS method (Small Baseline subset) - calculates accurate area displacements using a series of space images with small baselines. It includes the following stages: interferogram formation with further filtration and unwrapping, correction of orbital errors, first inversion, second inversion (atmospheric correction) and orthorrectification, geocoding of processing results. At the output we get a dot vector file, which is a map of average velocities of Earth surface displacements change. [4-5]

## 3 Results and disscution

In our work, we used the latter method since it allows us to reveal the dynamics of areal deformations of the earth's surface over a long period of time.

The object of the study was the northern part of the stockpile of the Kiyzassky open pit mine, on the territory of which a landslide occurred. For calculations, 11 radar images were selected from the Sentinel-1B spacecraft for the period from 04/12/2019 to 07/29/2019 with a shooting frequency of 12 days. This spacecraft performs imaging in C-band with 5x20 meters/pixel resolution (in Interferometric Wide Swath mode) and with VV and VH polarization.

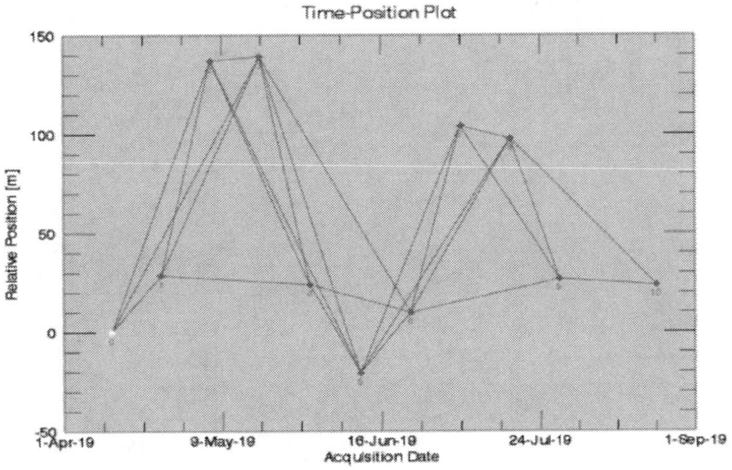


**Fig. 2.** Time-Position plot

Usually, the original radar image covers a large territory (several tens of thousands of square kilometers), so to reduce the time spent on processing a separate area of interest of about five square kilometers to the northeastern part of the "Kiizassky" section was allocated. To reduce the amount of time and intermediate processing data, VV polarization imagery was used. In addition, it has been experimentally determined that such polarization is better suited for determining displacements than VH. The whole processing was performed in the SARscape software package.

At the first stage, combinations of image pairs (a master image - an auxiliary image) and a network of connections (Fig. 2) were determined, which are used to form several differential interferograms.

It should be noted that those image pairs, which do not exceed the baseline threshold value (the default value is 500%), are suitable for processing. If the value is exceeded, pairs are excluded from further processing.

Then standard interferometric processing was performed with coregistration, filtering and phase unwrapping of image pairs selected at the first stage. Multilooking parameters equal to 1 in azimuth and 5 in range were used for image co-registration. To improve accuracy, we used the SRTM v4 digital elevation model with a spatial resolution of about 30 meters (1 arc second). Noise filtration was performed with the help of Goldstein filter. After that, the procedure of phase progression correction from the Earth ellipsoid was performed [6], where the topographic phase was removed from the obtained interferograms. The next step was two inversion stages. At the first stage, the resulting height and displacement rate are determined. The second step is to obtain offsets by dates, which are filtered to remove atmospheric phase components. Finally, geocoding of the results is performed with filtration by velocity and altitude to remove areas with high noise reading (areas with rich vegetation).

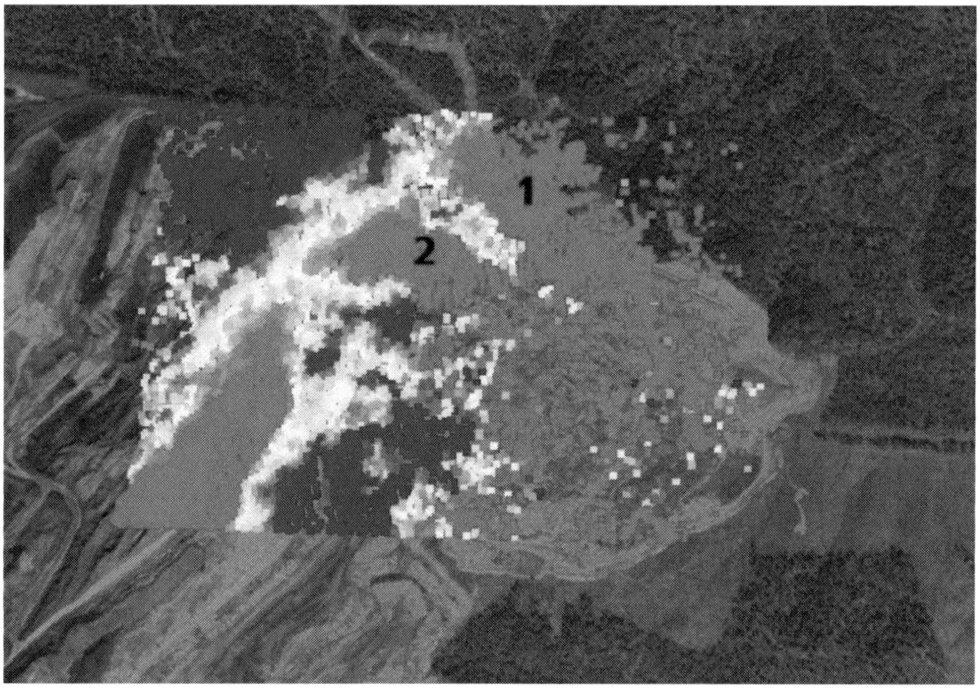


**Fig. 3**. Map of average velocities of vertical displacements.

During the processing of the initial images by the SBaS method, a map of the average rates of change in the vertical displacements of the earth's surface was built (Fig. 3). This map is a vector file with 4909 points. Each point contains a set of attributes, such as geographic coordinates, average displacement rate value, etc. Points with negative vertical displacements are highlighted in blue, which indicate subsidence of the earth's surface, and in red, elevations of the earth's surface.

On the map, figure 2 indicates the stockpile of the Kiizassky section, figure 1 - a section filled with rocks formed after the landslide. As can be seen from Figure 2, a dip is detected between sections 1 and 2, which means that rocks are coming off the side of the stockpile. In section 1, positive vertical displacements associated with an increase in the surface height due to the descent of rocks from the side and their further collision with the hillside are detected (Fig. 1).

Separately, it should be noted that the average displacement rates for the period from May 30, 2019 to June 11, 2019 (the landslide occurred on the night of June 7) increased threefold. The values of the average displacement rates over the observation period ranged from -170 mm / year to +200 mm / year.

Thus, the use of radar data from Sentinel-1 spacecraft is expedient for landslide monitoring, especially in areas with intensive mining operations. The obtained results may become the basis for the development of monitoring systems and landslide prevention activities.

The reported study was funded by RFBR and Kemerovo region, project number 20-47-420002.

## References

1. D. Raucoules, C. Colesanti, C. Carnec Use of SAR interferometry for detecting and assessing ground subsidence. Comptes Rendus Geosci, **339**, 289–302, (2007)
2. Stali izvestny prichiny CHP na razreze «Kijzasskij» v Kuzbasse https://gazeta.a42.ru/lenta/news/57608_stali-izvestny-prichiny-chp-na-razreze-kiizasskii-v-kuzbasse. Cited January 20, 2020.
3. Opolzen' na razreze Kijzasskij snyos shest' opor LEP, zavalil les i reku Kijzas. http://kuzpress.ru/incident/09-06-2019/68082.html. Cited January 20, 2020.
4. Kantemirov YU.I. Kratkiye teoreticheskiye osnovy radarnoy interferometrii yeye mnogoprokhodnykh variatsiy PS i SBaS // Geomatika. -2012.-№1.-S. 22-26
5. Rosen P A, Hensley S, Joughin I R, Li F K, Madsen SN, Rodrıguez E, Goldstein RM Synthetic aperture radar interferometry. Proc IEEE 88(3), 333–382, (2000)
6. Sputnikovyye tekhnologii v geofizike. http://media.ls.urfu.ru/566/1545/3670 Cited October 10, 2020.

# Spatiotemporal analysis of the land surface temperature distribution over the territory of Novosibirsk city based on Landsat data

*Elena A.* Mamash[1,*], *Igor A.* Pestunov[1], and *Dmitri L.* Chubarov[1]

[1]Federal Research Center for Information and Computational Technologies, Novosibirsk, Russia

**Abstract.** The paper discusses the applicability of Landsat 8 data for analyzing the land surface temperature distribution over the territory of Novosibirsk. The satellite data is compared with the data from ground meteorological stations. Using cloud-based systems and methods for processing time series of satellite data, a composite image of the temperature field for the territory of Novosibirsk is built; trends and anomalies in the temperature distribution in the urban area are studied. The results could be applied in urban development analysis and management of the city territory.

Availability of remote sensing data of low, moderate and high spatial resolution gave rise to development of tools for its online processing. Probably the most widely used such tool is Google Earth Engine, other examples include VEGA-Science developed at the Space Research Institute of Russian Academy of Science [1]. Federal Research Center for Information and Computational Technologies maintains an archive of satellite remote sensing data accumulated since 2007 and an information system for online retrieval and processing of satellite data time series [2].

Transition from individual satellite images to time series makes it possible not only to follow the dynamics of change but to compute statistical characteristics of different environmental characteristics such as land surface temperature (LST) of a certain region. This in turn permits further statistical processing such as removal of outliers and locating persistent anomalies.

Land surface temperature retrieved from Landsat data is used extensively for many applications including observation of urban heat islands, locating persistent anomalies, and monitoring urban sprawl. Examples include studies of urban heat islands of cities located in moderate climate zones such as Moscow and Nizhny Novgorod, as well as largest cities located above the Polar Circle, the cities of Murmansk, Norilsk, Vorkuta, Apatity, Salekhard, Noviy Urengoy, Nadym [3–7]. Detection of changes of the urban landscape of Krasnoyarsk was performed using the land surface temperature data retrieved from thermal infrared satellite imagery [8]. We remark that the authors of the papers cited above use individual satellite scenes for mapping and use pairs of images for the task of change detection. We hypothesize that such methods might be sensitive to the choice of images.

The uncertainty in satellite measurements of land surface temperature can be quantified by comparing LST retrievals with the data from ground weather stations [9].

*e-mail: elenamamash@gmail.com

To retrieve land surface temperature from Landsat data we apply the single channel algorithm [10, 11].

$$T_{LST} = \frac{T_B}{\left(1 + \lambda \frac{T_B}{c_2}\right) \cdot \ln \varepsilon} - 273.15, \qquad (1)$$

where $T_B$ denotes the brightness temperature (K) determined from observed radiation intensity in a satellite spectral channel, $\lambda$ - the central wavelength of the satellite spectral channel, for Landsat Band 10 it equals 10.8 μm, $c_2 = h \cdot c / k_B = 1.4388 \times 10^{-2}$ m K = 14 388 μm K is the second Planck radiation constant expressed as the product of Planck constant $h = 6.626 \times 10^{-34}$ J s; $k_B = 1.38 \times 10^{-23}$ J K$^{-1}$ the Boltzmann constant; $c = 2.998 \times 10^8$ m s$^{-1}$ is the speed of light; and $\varepsilon$ is the surface emissivity.

Several approaches to estimating surface emissivity can be found in the literature. The most commonly used are (1) $\varepsilon$ is considered constant usually equal to 0.98; (2) land use classification is performed and each class is assigned a specific emissivity coefficient (3) the emissivity coefficient is estimated from NDVI [12]. In this study we exclusively use the emissivity coefficient estimated from NDVI.

We compared the temperature values obtained with the Equation 1 with emissivity estimated from NDVI and with emissivity set constant at 0.98 and encountered differences that did not exceed a tenth of a degree. Therefore we conclude that the temperature is not particularly sensitive to this parameter.

When computing temporal composites we applied masks to exclude clouds and cloud shadows extracted from Landsat QA band.

To perform the computation and compose the multitemporal images we used Google Earth Engine. This instrument provides access to Landsat 8 data processed with atmospheric correction[1].

Landsat 8 thermal bands have spatial resolution of approximately 100 meters however in Google Earth Engine this data is resampled to 30 meters which makes this data particularly suitable for use at an urban scale.

We compare the land surface temperature maps with the land use data obtained from NextGIS[2] and map data from Google Maps and Bing Maps.

To validate the temperature data retrieved from Landsat we used the measurements from the five weather stations located in Novosibirsk. It is expected that the air temperature at 2 meter height can differ from the surface temperature, so we were looking not for absolute equality between the time series but for a significant correlation between the two. We retrieved archived weather data for the period from 2013 to 2019 from RP5 website[3]. The number of usable satellite measurements at the locations of the weather stations differs from station to station and from month to month yet there are common trends and the correlation was over 90% in all cases.

We used the available satellite data to perform a qualitative assessment of the land surface temperature field and to determine patterns in the distribution of surface temperature in Novosibirsk and adjacent areas, and to identify persistent thermal anomalies. Locating such thermal anomalies may be useful for environmental management and for the evaluation of the impact of industrial zones.

Figure 1a shows the composite image based on the series of Landsat 8 images for the periods without snow (from May to September in years 2013–2019). The next Figure 1b shows the same image with land use data imposed over it. On Figure 1c the previous image

[1] https://developers.google.com/earth-engine/datasets/catalog/landsat
[2] https://data.nextgis.com/ru/?lvl=regions&country=ru
[3] https://rp5.ru

is shown with the layer of roads and railways. The same process was applied to the data collected over the periods with snow (from November to March, 2013–2019)

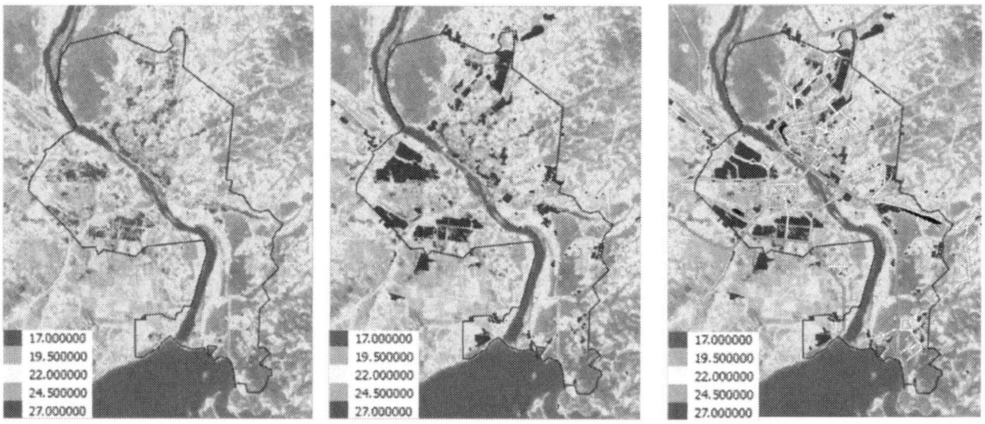


**Figure 1.** Temperature maps of Novosibirsk: (a) The original thermal image composite (b) composite with overlayed industrial zones (c) composite with industrial zones and road and railway networks overlayed

From Figure 1 one can see that the zones with above average surface temperature mostly correspond to industrial locations and the road network. In future we plan to perform a quantitative analysis of this link to determine the rate of growths of industrial zones.

We also studied the link between the temperature distribution and land use categories in Novosibirsk. The histogram and the temperature intervals corresponding to different categories are shown in Figure 2.

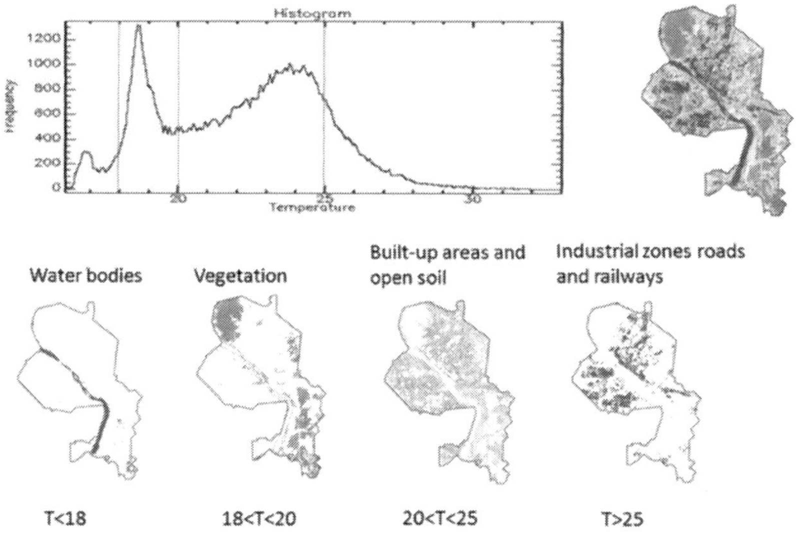


**Figure 2.** A histogram of average temperatures in Novosibirsk and the schematic maps of the land use categories corresponding to the modes of the histogram

**Table 1.** Main statistical characteristics by year

| Year | Scenes used | Min | Max | Mean | Stdev |
|---|---|---|---|---|---|
| 2019 | 51 | 14.03 | 34.45 | 21.86 | 3.04 |
| 2018 | 46 | 13.65 | 35.49 | 21.91 | 2.70 |
| 2017 | 50 | 4.74 | 40.58 | 22.82 | 3.40 |
| 2016 | 53 | 15.30 | 35.11 | 22.11 | 2.97 |
| 2015 | 43 | 13.71 | 37.04 | 21.76 | 3.40 |
| 2014 | 44 | 5.67 | 36.10 | 22.66 | 3.27 |

Empirically we identified the four main land use categories within the city:

1. water bodies,
2. vegetation,
3. built-up areas and open soil,
4. industrial zones roads and railways

In future work we intend to quantify the reliability of classification into the 4 main categories based on land surface temperature. This will inform us of the practicality of using surface temperature as an additional feature for land use classification.

The temperature maps show that while the temperature distribution during the periods without snow within the boundaries of the city may change year upon year the average temperature remains relatively constant (see Figure 3, Table 1).

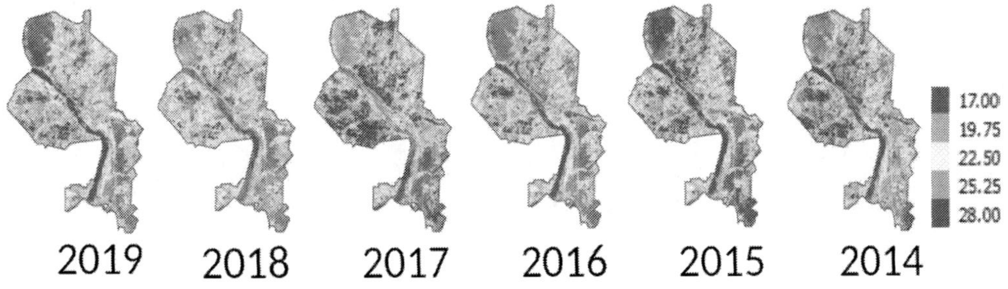


**Figure 3.** Spatial distribution of average temperatures during the periods without snow in Novosibirsk in 2014–2019

The results presented in this report include a method to produce land surface temperature maps based on the atmospherically corrected Landsat 8 satellite data. We confirmed that there is a strong correlation between the land surface temperature time series and air temperatures measured at weather stations at 2 meter height. Therefore land surface temperature can be used as a reliable source of information on the condition of the land cover of urban territories.

We also show that online instruments for processing of Landsat 8 time series can be used to identify spatial patterns in the land surface temperature distribution while the land surface temperature data can serve as an additional feature for classification of urban landscapes.

## References

[1] A.A. Proshin, E.A. Lupian, I.V. Balashov, A.V. Kashnitskii, and M.A. Burtsev, Current Problems in Remote Sensing of the Earth from Space, **13(3)**, 9–27 (2016)

[2] Yu.I. Shokin, I.A. Pestunov, D.L. Chubarov, V.A. Kikhtenko, E.A. Mamash, N.N. Dobretsov, P.V. Voronina, and Yu.N. Sinyavsky, CEUR Workshop Proceedings, 2534, 229–240 (2019)

[3] E.A. Baldina and M.Y. Grishchenko, South-Eastern European Journal of Earth Observation and Geomatics, **3(2S)**, 415–418 (2014)

[4] M.I. Varentsov, P.I. Konstantinov, T.E. Samsonov, and I.A. Repina, Current Problems in Remote Sensing of the Earth from Space, **11(4)**, 329–337 (2015)

[5] M.Y. Grishchenko, M.I. Varentsov, and P. G. Mikhailyukova, Proceedings of the International Conference on Regional Problems of Remote Sensing of the Earth, 217–219 (2019)

[6] M.I. Varentsov, M.Y. Grishchenko, and H. Wouters, Geography, Environment, Sustainability, **12(4)**, 74–95 (2019)

[7] M.Y. Grischenko and J.V. Yermilova, Geodesy and Cartography, **79(3)**, 23–34 (2018)

[8] A.A. Gosteva, A.K. Matuzko and O.E. Yakubailik, Proceedings of the All-Russian Conference "Spatial Data Processing for Monitoring of Natural and Anthropogenic Processes" (SDM-2019), 387–392(2019)

[9] A.K. Matuzko, O.E. Yakubailik, Journal of Siberian Federal University, series Engineering and Technologies, **11(8)**, 934–945 (2018)

[10] J.C. Jiménez-Muñoz, J.A. Sobrino, D. Skoković, C. Mattar, and J. Cristóbal, IEEE Geoscience Remote Sensing Letters, **11**, 1840–1843 (2014)

[11] D. Parastatidis, Z. Mitraka, N. Chrysoulakis and M. Abrams, Remote Sensing, **9(12)**, 1208 (2014)

[12] A. Sekertekin, and S. Bonafoni, Remote Sensing, **12(2)**, 294 (2020)

# Geoecological agricultural lands of the Altai region of the Republic of Khakassia

*Ekaterina V.* Pavlova[1*], *Anastasiia I.* Volkova[1], *and Ekaterina A.* Demina[1]

[1]Institute of Natural Sciences and Mathematics, Katanova Khakass State University, 655000, Abakan, 90 Lenin avenue., Russia Federation

**Abstract.** The article discusses the issue of application geographic information systems for inventory agricultural land and assessment of their condition. A geoinformation project has been developed for conducting geoecological appraisals agricultural land of the Altai region of the Republic of Khakassia. The structure of agricultural land, the area determined fallow lands and an inventory of forest belts investigated territory has been studied based on analysis of the agricultural map of scale 1: 100,000, interpretation of Landsat 4-5, 7, 8 and Sentinel 1 and 2 images, and also ground based satellite research. To processes overgrowth tree and shrub vegetation the deposits subject were revealed. The research results showed that on the study area in the structure of agricultural lands 56860 hectares of land belong to the deposits located on different stages of recovery. The elm overgrowing processes covered about 28580 hectares of post-agrogenic lands.

The State program of the Republic of Khakassia «The development of agro-industrial complex of the Republic of Khakassia and the social sphere in the countryside» provides increasing in production and processing of main types of products crop production, development of small business in the countryside, as well as increasing productivity and sustainability of agricultural production and fertility soil by means of complex land reclamation in the face of climate change and natural anomalies [1]. The implementation of the assigned tasks primarily depends on turn from the efficiency of the use of agricultural land. Unfortunately, the all-Russian tendency to increase the share of unused arable lands (fallow lands) and lands with lost fertility, which is typical for The Republic of Khakassia, does not contribute to the solution of the above tasks.

Solving the problems of rational use of land resources requires relevant information of the actual ratio of agricultural land (arable land, hayfields, pastures, deposit, etc.) in agricultural landscapes, their features use, degradation processes (salinization, overgrowth, erosion, etc.), condition of shelter and pasture shelter belts, etc.

In this regard, the goal of our work is to create a geoinformational (GIS) project «The Geoecological Assessment agricultural land of the Altai region of the Republic of Khakassia» containing a series of maps reflecting the structure and current state of agricultural lands of the Altai region of the Republic of Khakassia.

* Corresponding author: eve21@yandex.ru

The Altai region of the Republic of Khakassia is located in its southern part within Koibalskaya steppe of the South Minusinsk depression. The relief is an elevated, gently sloping plain with separate low ridges of hills and small closed lake basins. The main part of the territory characterized by absolute relief marks from 260 to 580 m above sea level. The climate of the region is sharply continental. Duration of the period with temperature below 0°C averages 85 –110 days [2].

The amount of precipitation in May-July (the period of active vegetation of plants) varies from 170 to 210 mm, and in general, from 370 to 490 mm of precipitation falls per year. The duration of the frost-free period within the study area varies from 105 to 115 days. The dryness index based on a ratio the sum of the average daily temperatures above 10 ° C and the amount of precipitation for May - July is 1,9–1,7 (arid - insufficiently humidified) [3]. An essential feature of the climate is constant and quite strong winds. The dominant annual cycle is the south-westerly winds. Dust storms are often observed.

The main types of soils are chestnut, southern chernozems and ordinary, often chipped with bedrock outcrops. Characteristically widespread development of sandy and sifted sandy loam soils. In the lakeside depressions, depressions, in the floodplains of rivers, meadow-boggy, meadow, meadow chernozem or meadow chestnut soils. The alkali-saline soil are formed in conditions of sharp contrast water regime and alternating salinization and desalination.

According to geobotanical zoning of A.V. Kuminova (1970) the territory of the Altai region is a part of the Priabakan (Central-Khakass) geobotanical district of the Minusinsk Depression province [4].

Natural herbaceous vegetation preserved after the development of virgin lands by separate fragments are represented by small sod and large sod real steppes with their digression and crushed versions. By slopes uplands petrophytic true steppes are common and in the lakeside depressions halophytic vegetation. Woody shrub vegetation (willows, poplars) is usually confined to river valleys, birch groves are found in the eastern part of the region near the Yenisey river. On sandy soils Artificial pine forests are located in small tracts.

The following data are the result of many years of field research (2004-2020): the peculiarities of the use of agricultural land, the location of deposits and forest belts, their structure and preservation, the description of the stand in individual areas of fallows overgrown with wood shrub vegetation (species composition, age, trunk diameter and tree stand closeness). After the inventory the forest belts were divided into the following categories of preservation:

1 - not requiring restoration (100-70% of the total length forest belt);

2 - with preservation of more than 50% (69-50% of the total length of the forest belt);

3 - requiring restoration (49-10% of the total length of the forest belt);

4 - lost (9-0% of the total length of the forest belt).

With the help of ArcMap 10.1 the GIS project was created to conduct a geoecological assessment of agricultural land. That project included vector layers of farmland and forest belts and mosaics space images. The analysis of the following objects were made to create the GIS project: topographic maps (M 1: 100000), land use maps by regions (M 1: 100000, M 1: 50,000), public cadastral map of the Russian Federation, multispectral satellite images of the growing season (end of July - September) from 2009 to 2019 satellite images of Landsat 5, 7, 8 from the archive of the US Geological Survey (USGS, https://earthexplorer.usgs.gov) and Sentinel 1, 2A and 2B (URL:https://scihub.copernicus.eu), web services providing access to space high-resolution photography (ArcGis Online, Bing maps, Yandex maps, kosmosninki.ru) archival materials of satellite imagery from the interactive Google Earth Pro applications (spatial resolution - about 0,5 m). Based on 30 selected multispectral images of 2019 in the ArcMar 10.1 program, a catalog of space images was created, the georeferencing was corrected, the color rendition of the mosaic was corrected with alignment in tone (color balancing, histogram normalization). The boundaries of the spread of overgrowth and the area of the contours of fields affected by overgrowing with trees and shrubs were clarified with the help of the created mosaic of images from 2018-2019 from the end of July to September.

A vector line layer digitizing of the current location of forest belts was carried out according to field research data and actual satellite pictures. To assess the retrospective state of forest belts, we used agricultural maps of the Altai region.

The ArcMap working project settings are the following: coordinate system WGS 1984 UTM Zone 46N, Mercator / WGS84 projection. The project includes vector linear layers: rivers, roads of various categories, boundaries of districts, forest belts and areal vector layer of farmland. According to field research, web services and visual decoding of space images, data on the types of land and the degree of overgrowth of deposits were entered into attributive information of the areal vector layer of agricultural fields. In the linear layer of forest belts indicated the category of their safety, length, number of stripes and tree species. By the results of the analysis of the attribute information were calculated arable land, hayfields, pastures and post-agrogenic territories with different degree of overgrowth.

To assess the process of overgrowth with vegetation, the following stages of overgrowth were identified:

1. Zero stage of overgrowth - there are no overgrowing processes.

2. The first stage of overgrowth - undergrowth of tree species appears commensurate with the herbaceous layer with a projective cover of up to 50-100%, especially near the parent plantations (forest belts).

3. The second stage of overgrowth - thinning of the undergrowth begins due to biological characteristics of the elm (a large percentage of frosting of plants), the density of the stand is reduced to 0,3-0,5.

4. The third stage of overgrowth is further thinning of the stand. Completeness is very low $\leq 0,1$. The average height of the elm is at least 2 m (as a rule the age of the plantings is over 8 years). Young undergrowth reappears [5].

The area of the Altai Municipal District is 173,61 thousand ha of which agricultural land (163,85 thousand ha) occupy 94,4% of the total area of the district. The soils on the territory of the region are mainly of low and medium unstable fertility and are mainly suitable for fodder land and use as arable land with limited sowing of row crops. In this regard, most of the agricultural land is not involved in the cultivation of agricultural crops and is represented by deposits at various stages of recovery (mainly wormwood-cereal and fire-wheatgrass deposits) (fig. 1).

Arable land is cultivated, and crops are grown mainly in the eastern and south-eastern parts of the Altai region within the Kirovsky, the Novomikhaylovsky and the Novorossiysk village councils. It is here that soils with high sustainable fertility are located, which are suitable for growing all agricultural crops. In total, as arable land in 2018-2019 about 55 thousand hectares were used. On the territory of the Altai district, 56860 hectares of agricultural land (32,5% of the total area of the district) are deposits at different stages of restoration. They are used for mowing in some wetter years. The process of demutation that has taken place on fallow lands in recent decades is significantly changing and slowing down due to the spread of the squat elm (Ulmus pumila L.). This tree species quite resistant to various influences (arid conditions, pests, pruning, etc.), acts as the basis of most of the forest shelter belts in Khakassia, including those within the Altai region. Elm forest belts act as a mother stand and centers of occurrence and development of overgrowing processes.

 https://doi.org/10.1051/e3sconf/202022303012

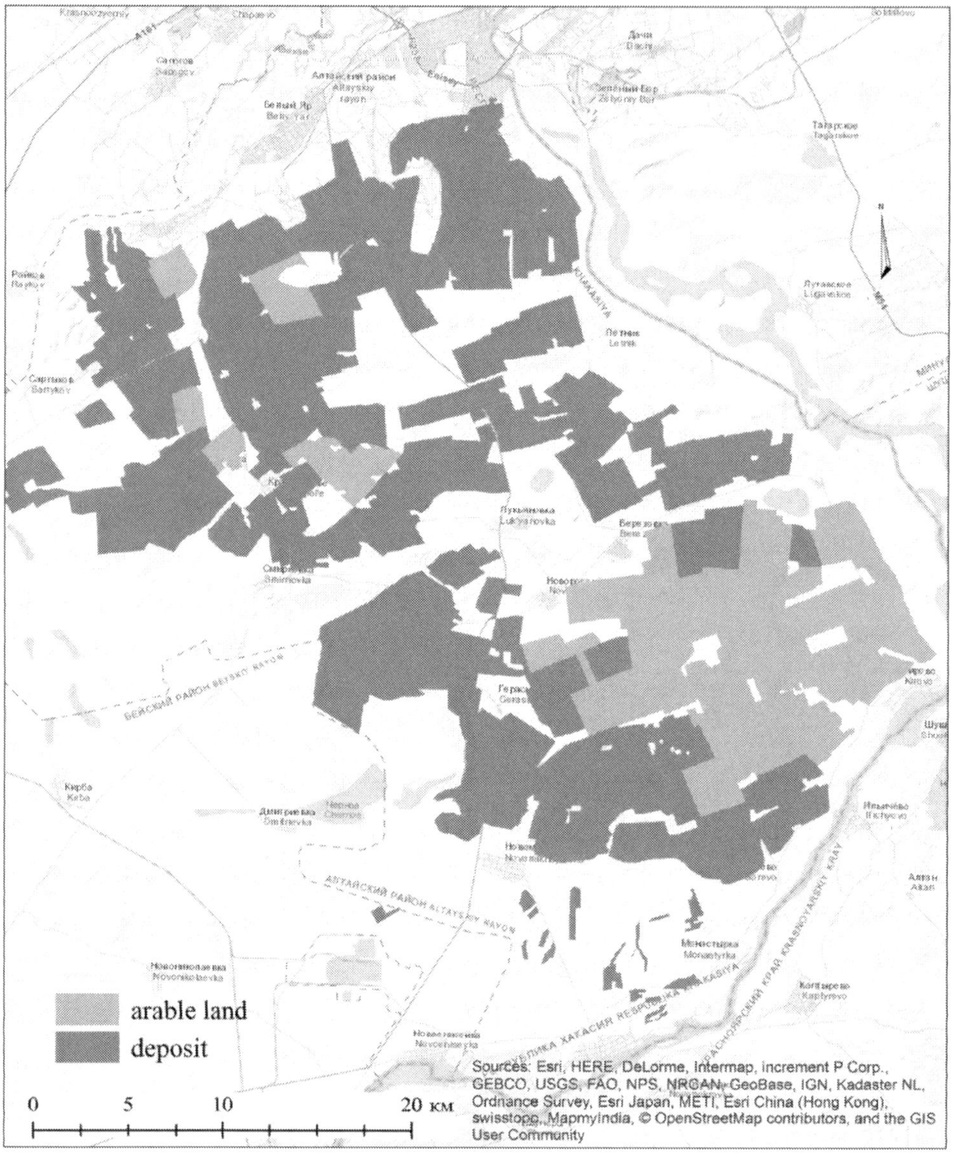


**Fig. 1.** Location of arable land and deposits in the Altai region.

It was revealed that in 2019 about 28580 hectares of post-agrogenic lands were covered by elm overgrowing processes. 7938 hectares of them are at the first stage of overgrowth, when undergrowth of tree species appears commensurate with the herbaceous layer with a projective cover of up to 50-100%, especially near the parent plantations (forest belts). Share of deposits, in the second stage of overgrowth is 49, 6% of the total area fallow lands (fig. 2).

 https://doi.org/10.1051/e3sconf/202022303012

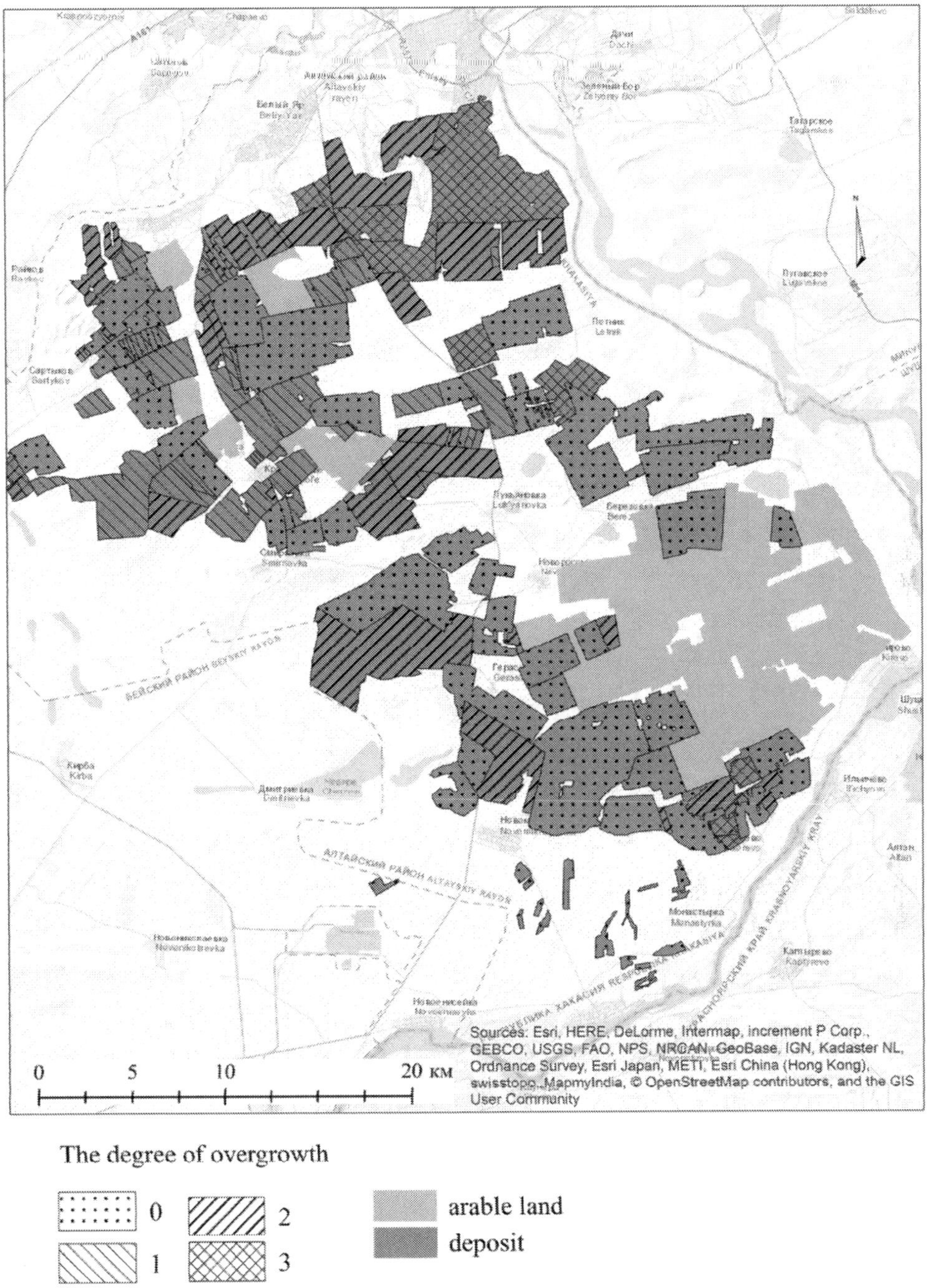


**Fig. 2.** Overgrowth of agricultural land by tree and shrub vegetation in the Altai region of Khakassia.

The title is set in bold 16-point Arial, justified. The first letter of the title should be capitalised with the rest in lower case. You should leave 22 mm of space above the title and 6 mm after the title.

The developed GIS project will provide land users with an up-to-date information on the structure of agricultural land and their condition can be used as a basis for drawing up a program for the use of lands with greater economic efficiency and with less environmental risk. The obtained results provide the basis for the development of an optimal placement of land and crops, the basis for the assessment of favorable and unfavorable consequences of environmental protection measures taken on the ecosystem as a whole and its individual components, making operational decisions on their adjustment depending on changing external conditions.

## References

1. The decree of the Government of the Republic of Khakassia of November 19, 2012 no. 781 «On approval state program of the Republic of Khakassia «The Development of agro-industrial complex of the Republic of Khakassia and the social sphere in the countryside» (as amended by 31 July 2020). URL: http://docs.cntd.ru/document/453366265.
2. A. I. Volkova, E. V. Pavlova, E. A. Demina. RPERS 2020. pp. 177- 179 (2020).
3. South of the Krasnoyarsk Territory. Series of maps for agricultural planning, 40 p. (1977).
4. A. V. Kuminova, Y. M. Maskaev, G. A. Zvereva, Vegetation cover of Khakassia (Nauka Pbs., Novosibirsk, 1976) in Russian , 421 p. (1976).
5. E. V. Pavlova, A. I. Volkova, E. A. Demina. E3S Web of Conferences, 149, 03006 (2020).

# Automated multi-classifier recognition of atmospheric turbulent structures obtained by Doppler lidar

*Anton* Sokolov[1,*], *Egor* Dmitriev[2], *Ioannis* Cheliotis[1], *Hervé* Delbarre[1], *Elsa* Dieudonne[1], *Patrick* Augustin[1], and *Marc* Fourmentin[1]

[1]University of Littoral Cote d'Opale, Laboratory for Physico-Chemistry of the Atmosphere, Dunkerque, France
[2]Institute of Numerical Mathematics of Russian Academy of Sciences, Moscow, Russia

**Abstract.** We present algorithms and results of automated processing of LiDAR measurements obtained during VEGILOT measuring campaign in Paris in autumn 2014 in order to study horizontal turbulent atmospheric regimes on urban scales. To process images obtained by horizontal atmospheric scanning using Doppler LiDAR, the method is proposed based on texture analysis and classification using supervised machine learning algorithms. The results of the parallel classification by various classifiers were combined using the majority voting strategy. The obtained estimates of accuracy demonstrate the efficiency of the proposed method for solving the problem of remote sensing of regional-scale turbulent patterns in the atmosphere.

## 1 Introduction

Atmospheric turbulence is a key meteorological characteristic, responsible for the dispersion of the air pollution and the cloud formation. When wind speed measurements are available with sufficient spatial and temporal resolution, the estimation of turbulence parameters becomes possible. As an example, the ground-based Doppler wind lidar Leosphere WLS100 allows remote measurements of radial wind speed profiles with the temporal resolution of about one profile by second and the spatial resolution of 50 meters.

The measurement database used in this study is based on data of VEGILOT measurement campaign [1], which was held in Paris in September – October 2014. This campaign was aimed at studying urban atmospheric dynamics, air pollution and turbulent regimes from lidar measurements. More information about VEGILOT campaign could be found in [1].

Horizontal wind turbulent patterns was calculated based on horizontal radial wind scans (see [2, 3]). Typical examples of these images are presented at Fig.1. Three classes of local atmospheric patterns were introduced: Thermals, Rolls and Streaks. An additional fourth class 'Others' contains patterns that could not be classified as these three turbulence types.

* Corresponding author: anton.sokolov@univ-littoral.fr

Each type of turbulence structure forms a specific cloud pattern and could be observed on satellite images.

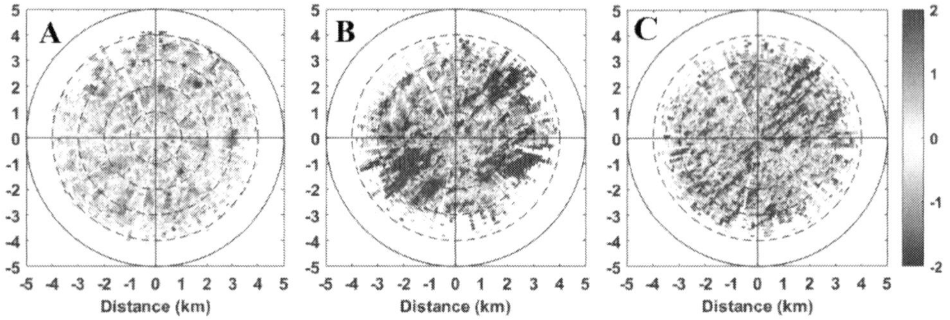


**Fig. 1.** Examples of turbulent structure patterns: A) Thermals observed on October 10 at 1111 UTC, B) Rolls observed on September 11 at 1100 UTC and C) Streaks observed on September 11 at 1957 UTC.

## 2 Description of Algorithms

A few thousands of images was obtained for the two-month campaign. The following supervise machine learning SML algorithms was applied to classify local atmospheric patterns:
• Parzen – Rozenblatt window (PRW);
• K-nearest-neighbors (KNN), a few number of neighbors were tested: K=1, K=3, and K=5;
• Error-correcting output codes with support vector machine (SVM);
• Quadratic discriminant analysis (QDA).

An expert classified 150 patterns to construct the training set [2]. The in-situ meteorological and satellite data were used in addition to lidar measurements. Since SML algorithms require numerical values characterizing patterns, Haralick texture features were applied for image classification [4]. Below we describe briefly how the features were calculated, see [2] for more details.

The following four statistics were calculated: Contrast, Homogeneity, Correlation and Energy. Each of those characteristics was calculated at different distances and directions of adjacency. The neighbors ranging from first until 30th for each statistic allow generating 30*4 = 120 angular functions [2, 4]. Each of these functions was characterized by the following three properties: amplitude (maximum value minus minimum value), integral, symmetry.

Three additional features were added to those 30*4*3 = 360 textural features, namely the time of the scan, the average wind speed and the cosine fit error [2, 3]. Thus, 363 features (predictors) were calculated for each turbulent image.

To avoid the curse of dimensionality, the cross-validation stepwise forward selection method was implemented, since the dimension of feature vector largely exceeds the number of patterns in the learning ensemble. The number of optimal features selected [2, 5] vary between 2 and 20.

The SML classifiers were combined in one multiple classification (MC) algorithm using the majority voting strategy [6].

## 3 Results and Discussion

The total overall accuracy (TA) score for each classification technique was presented in the table 1. It shows the percentage of correctly classified images based on cross validation.

We can see that among applied standard SML algorithms the QDA has the best TA performance, following by SVM and KNN. The PRW algorithm has the lowest TA. It shows that SML algorithms, that produce more complex decision boundaries are less accurate for the given classification problem. We can also note the significant TA improvement for the proposed MC algorithm.

**Table 1.** Total overall accuracy for applied SML algorithms.

| SML Algorithm | PRW | 1NN | 3NN | 5NN | SVM | QDA | MC |
|---|---|---|---|---|---|---|---|
| TA, % | 82.7 | 90 | 88.7 | 89.3 | 90.7 | 92 | 94.7 |

A confusion matrix for MC is presented in Fig. 2A. All types of turbulent patterns are identified with good accuracy. After the learning step, in the classification step, the MC algorithm was applied to a complete lidar dataset (test set) of 4557 patterns. Streaks was detected in 23% of cases, Rolls in 10% and Thermals in 17%.

In Fig. 2B, the distribution of turbulence types by the time UTC is shown. As expected, Streaks are generally observed during the nighttime, while Thermals and Rolls are detected in the daytime (see [2] for more details).

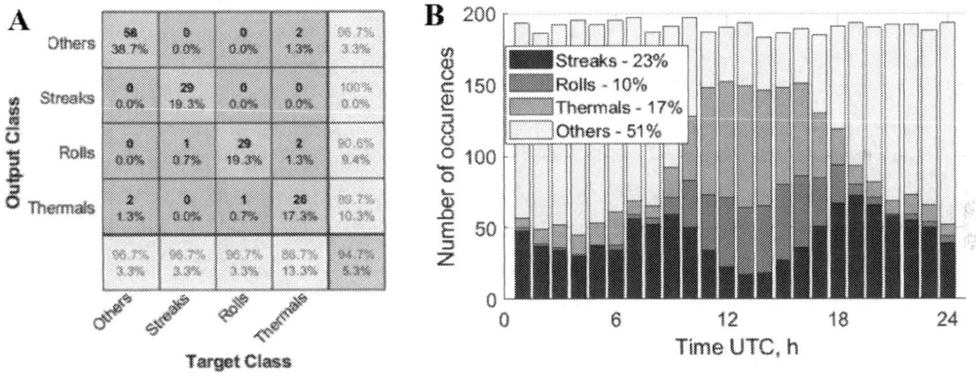


**Fig. 2.** A) Confusion Matrix for MC algorithm. B) Histogram of resulting distribution of turbulent classes by time.

## 4 Conclusions and Perspectives

On the basis of the results of large lidar dataset processing, we can conclude that the proposed method allows efficient solving the problem of turbulent regimes classification. The comparison of SML algorithms shows that the accuracy of relatively simple QDA is better than accuracy of other SML algorithms that construct more complicated decision boundaries (PRW, KNN and SVM). An important increase of the total accuracy could be achieved by combining a few SML algorithms in MC system. This technique was successfully applied for the study of atmospheric turbulence in Paris region in the autumn.

In the future studies, we are going to optimize the procedure of classifiers' combination and to increase the number of Haralick texture features. The proposed algorithm will be applied to characterize the atmospheric turbulence in coastal areas.

This work is a contribution to the CPER research project IRenE and CLIMIBIO. The authors thank the French Ministère de l'Enseignement Supérieur et de la Recherche, the Hauts-de-France Region and the European Funds for Regional Economic Development for their financial support to this project. The work is supported by the CaPPA project. The CaPPA project (Chemical and Physical Properties of the Atmosphere) is funded by the French National Research Agency (ANR) through the PIA (Programme d'Investissement d'Avenir) under contract "ANR- 11-LABX-0005-01" and by the Regional Council Nord-Pas de Calais and the European Funds for Regional Economic Development (FEDER).
The work was supported by the Moscow Center for Fundamental and Applied Mathematics (Agreement 075-15-2019-1624 with the Ministry of Education and Science of the Russian Federation) and Russian Foundation for Basic Research (RFBR, project № 20-07-00370).
The aerosol lidar belongs to the air quality observatory, QUALAIR (http://qualair.aero.jussieu.fr), located at the Université Pierre et Marie Curie.
Experiments presented in this paper were carried out using the CALCULCO computing platform, supported by SCoSI/ULCO (Service COmmun du Système d'Information de l'Université du Littoral Côte d'Opale).

## References

1. A. Klein, F. Ravetta, J.L. Thomas, G. Ancellet, P. Augustin, R. Wilson, E. Dieudonné, M. Fourmentin, H. Delbarre, J. Pelon. Influence of vertical mixing and nighttime transport on surface ozone variability in the morning in Paris and the surrounding region. Atmos. Environ. **197**, 92-102. (2019) https://doi.org/10.1016/j.atmosenv.2018.10.009
2. I. Cheliotis, E. Dieudonné, H. Delbarre, A. Sokolov, E. Dmitriev, P. Augustin, M. Fourmentin. Detecting turbulent structures on single Doppler lidar large datasets: an automated classification method for horizontal scans, accepted for publication to Atmos. Meas. Tech. journal, 2020. https://doi.org/10.5194/amt-2020-82
3. W.L. Eberhard, R.E. Cupp, K.R. Healy. Doppler Lidar Measurement of Profiles of Turbulence and Momentum Flux. Journal of Atmospheric and Oceanic Technology, **6(5)**, 809-819. (1989) https://doi.org/10.1175/1520-0426(1989)
4. R.M. Haralick, I. Dinstein, K. Shanmugam. Textural Features for Image Classification. IEEE Transactions on Systems, Man and Cybernetics, **SMC-3(6),** 610-621. (1973). https://doi.org/10.1109/TSMC.1973.4309314
5. A. Sokolov, E. Dmitriev, C. Gengembre, H. Delbarre. Automated classification of regional meteorological events in a coastal area using in-situ measurements. J. Atmos. Ocean. Technol. JTECH-D-19-0120.1. (2020) https://doi.org/10.1175/JTECH-D-19-0120.1
6. R. Ranawana and V. Palade. 'Multi-Classifier Systems: Review and a Roadmap for Developers'. IJHIS, **3, 1**, 35-61 (2006).

# Analysis of Long-Term Gravitational and Seismic Measurements in the Pamir - Tien-Shan Region

*Leonid Sverdlik*[1,*], *Ekaterina Lazareva*[1]

[1]Research Station of the Russian Academy of Sciences in Bishkek, Kyrgyzstan

**Abstract.** Long-term variations (2003–2015) of the Equivalent Water Height (EWH) above the geoid's contour obtained from GRACE satellite measurements are analyzed to establish the relationship between changes in the gravitational field with the stress-strain state of the geological environment and the seismic process in Central Asia (Pamir and Tien-Shan regions). The earthquakes gravitational effects study, based on the use of various GRACE models allows us to distinguish the component of the gravity field, which may be related to seismic activity. The results of temporal and spatial gravitational anomalies identification, the probable causes of which were tectonic factors associated with the redistribution of masses in the lithosphere as a result of large (M>5.5) regional earthquakes, are presented.

## 1. Introduction

The study of spatial and temporal changes in the gravity field was performed in one of the most seismically active areas of Central Asia, located in the junction of two major mountain systems – Pamir and Tien-Shan. High seismicity of the region is conditioned by intense geodynamic interaction of several large lithosphere plates. It is assumed that regional earthquakes focus generation is direct consequence of submeridional compression high tensions, the energy source of which is pressure exerted by the Hindostan plate from the South [1]. Over several last centuries, more than ten earthquakes of M>7.0 have occurred in the Pamir and Tien-Shan regions [2]. However, the complex and rugged terrain of the study area make it impossible to conduct ground-based studies of geodynamic and deformation processes based on traditional geophysical and geodetic measurements.

Satellite gravimetric data obtained during the GRACE experiment (*Gravity Recovery and Climate Experiment*), are widely used to study various geophysical phenomena and processes. The GRACE data archive contains satellite measurements for 15 years (2002-2017). Monthly gravity models in the form of a breakdown by spherical functions are calculated by several research centers in Switzerland (AIUB), Germany (GFZ), USA (CSR and JPL) and France (CNES/GRGS) [3], that generate monthly gravity fields uniformly covering the Earth surface. The high precision of the GRACE system, being $1\times10^{-10}$ m/s$^2$

* Corresponding author: l.sverdlik@mail.ru

(tens of micrometers), determines sensitivity of this satellite gravimetric means to integral mass redistribution, that allows to explore not only dynamics of oceans, volume of glaciers, seasonal and interannual hydrologic changes, but geodynamic process, accompanied by density fluctuations in the Earth's crust and mantle in the process of major earthquakes as well. Changes of the gravity field associated with several strongest earthquakes, e.g. Sumatra (2004), Chili (2010) and Japan (2011) [4] are examples of successful detection of post-seismic effects. Sudden changes of gravity field values, representing negative changes of about 4 μGal, were registered in the month when the events occur.

This paper analyzes long-term regional data from satellite measurements (GRACE) in order to identify coseismic gravitational deformations and establish the relationship of gravitational field anomalies with the seismic process in the Pamir and Tien-Shan regions.

## 2. Gravitational And Seismic Measurements Data

The initial data were monthly gravitational field models of the Geophysical institute in Potsdam (GFZ-RL06) [5], Centre National d'Etudes Spatiales / Groupe de Recherches de Géodésie Spatiale, Toulouse, France (CNES/GRGS-RL05) and the Astronomical Institute of the University of Bern (AIUB-RL02) [6]. Researches of spatial-temporal changes in the gravity field were performed in the most seismically active region of the Central Asia, which is marked with a dashed line in the map of earthquakes focuses distribution, occurred in the region limited by the following coordinates: 37.0–44.0°N, 70.0–79.0°E, in 2003–2015. (Fig. 1).

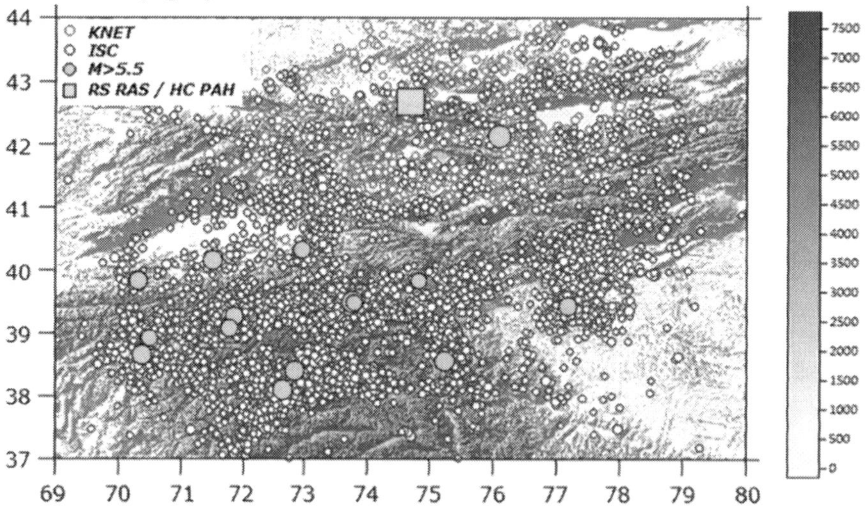


Fig. 1. Location of M≥2.0 earthquake focuses (KNET), M≥3.0 and M≥5.5 (ISC)

Information about earthquakes was obtained according to data of the KNET seismic stations regional network (M≥2.0) and on the ISC interactive service (M≥3.0) [7]. To characterize the seismic regime, the slope parameter of the repeatability graph (*b-value*) was used, which reflects the distribution of the earthquakes number by magnitude (M). The *b-value* parameter was estimated by the least squares method for the cumulative representation of the initial seismic data in a sliding time window of 6 months. The *b-value* parameter depends on the ratio between the number of strong and weak earthquakes and is directly related to the distribution of stresses in the Earth's crust [8].

## 3. Results and Discussion

### 3.1. Anomalous variations in the gravitational field associated with strong earthquakes

To identify gravity field anomalies associated with seismic activity, GRACE data was analyzed in the form of solutions from two analytical centers AIUB-RL02 and CNES/GRGS-RL05 (Fig. 2a). These variations reflect integral changes relative to the average gravitational field (converted to the equivalent water level above the geoid's contour, EWH) in the area of M≥5.5earthquake concentration (Fig. 1). The results presented in Fig. 2a indicate a seasonal component as a dominant part of EWH time series. They also show that the peak values of seasonal components of changes in the gravitational field are observed in spring and autumn.

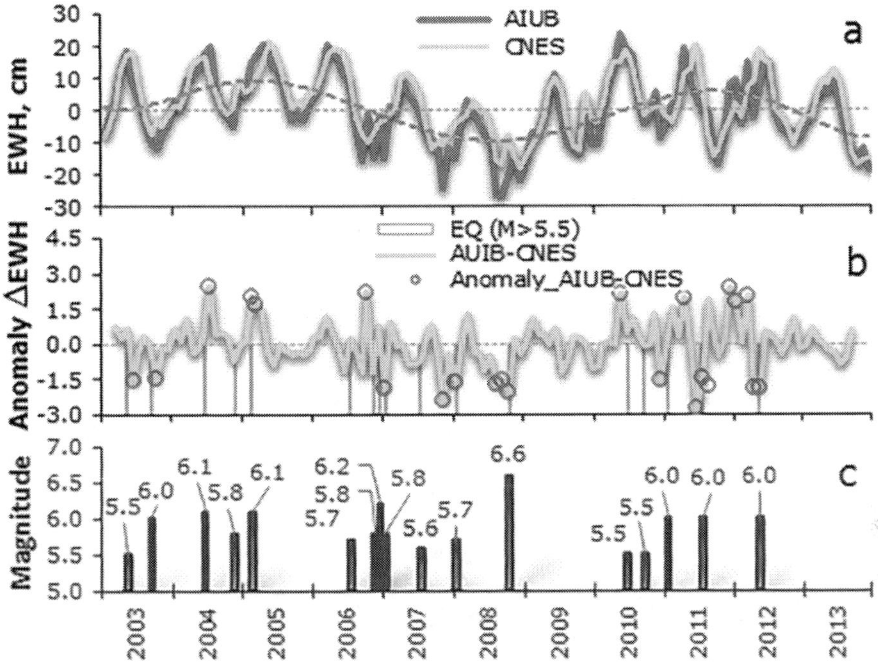


Fig. 2. Time series of gravitational field variations AIUB-RL02 and CNES/GRGS-RL05 (a), the gravitational field high-frequency component anomalies (b) and the sequence of M≥5.5 earthquake magnitudes (c) in 2003-2013

It should be noted that the models of the gravitational field provided by these research centers are obtained using different approaches to satellite observations processing, considering different numbers of spherical harmonics coefficients, and differ in the filtering method. Accordingly, in spite of obvious consistency between two GRACE time series (Fig. 2a), that have similar phases, a close amplitude and identical trend, there are some significant differences. AIUB-RL02 models have more high-frequency variations, than those of CNES/GRGS-RL05. Taking into consideration that the gravity field changes under the influence of various geophysical processes, most significant of which are conditioned by natural seasonal periodicity and interannual hydrologic changes on the Earth surface, the above specifics of the time series allow to identify the gravity field component that may be associated with seismic activity. Calculation of difference between time series of AIUB-RL02 and CNES/GRGS-RL05 (ΔEWH) allowed to exclude seasonal and trend changes while not applying traditional methods of seasonal decomposition:

$$\Delta EWH(t) = EWH_{AIUB}(t) - EWH_{CNES/GRGS}(t) \quad (1)$$

As the AIUB-RL02 time series do not cover the whole available GRACE measurement set and contain a lesser number of monthly models than CNES/GRGS provides, comparison results for the 2003–2013 period are presented. Within the assumption of normal distribution of $\Delta EWH$ with an average value of μ(ΔEWH) and the standard deviation of σ(ΔEWH) the anomality of the selected gravity field irregular constituent of ΔEWH (Fig. 2b) was determined according to the condition of δEWH values overrunning of the ±1.34σ interval limits in accordance with the expression presented in the [9] paper. In this case the level of confidence was 82%:

$$\delta EWH(t) = [\Delta EWH_i(t) - \mu(\Delta EWH)] / \sigma(\Delta EWH) \qquad (2)$$

Analyzing the EWH variations, it can be seen that sharp anomalous changes of δEWH are simultaneous with most major earthquakes (M≥5.5), occurred in the territory under consideration, or precede them by a month (Fig. 2c). It allows to conclude that the observed anomalies of the gravity field were of seismic origin. The differently directed nature of δEWH is probably conditioned by pre- and coseismic mass deformations and redistribution, emerging in a density inhomogeneous geologic environment due to formation of areas of higher and lower stresses.

Preliminary analysis of the results showed obvious relation between δEWH gravity field anomalies and M≥5.0 earthquake magnitude. For negative anomalies we obtained a higher correlation coefficient equal to ~0.55.

### 3.2. Long-Period Changes in the Gravitational Field

Despite the limited duration of the GRACE mission, the results of satellite observations allow us to analyze variations in the gravitational field, the period of which exceeds 12 months. Fig. 3a shows the EWH variations smoothed with a moving average over a time window of 12 months with a time shift of 1 month and a polynomial of the 6th degree.

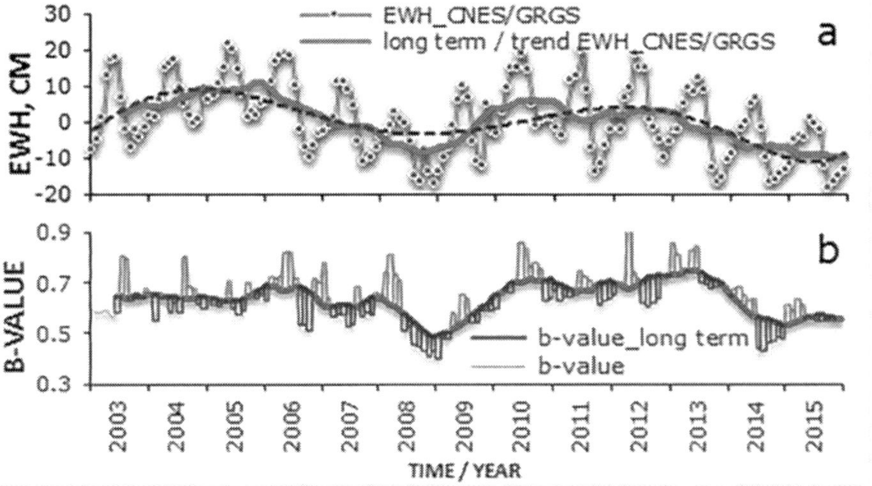


Fig. 3. Long-period / trend changes of EWH (a) and the b-value parameter (b) in 2003-2015

The feature of the identified long-period EWH variations is a significant unevenness of the change over time with a pronounced decreasing trend in 2007–2008 and 2014–2015. These time intervals were characterized by maximum decreases in the density of the geological environment, which is probably due to a high level of seismicity and, consequently, changes in the stress-strain state of the geological environment. Activation of the seismic mode corresponds to a decrease in the slope of the repeatability graph to

~0.45–0.50 (Fig. 3b). Periods of relatively weak seismicity without strong M≥6.0 earthquakes in the study area, such as in 2010, were characterized by higher *b-value* parameters of ~0.65–0.70. The minima in the equivalent water level changes and slopes of the earthquake recurrence graphs (*b-value* parameter) corresponded to periods of regional seismicity enhancement and coincided with the major Nurinsk earthquake (magnitude M=6.6; October 10, 2008) and the seismic event with M=6.5 magnitude that occurred in December 07, 2015.

### 3.3 The Spatial Distribution of the Gravitational Field Anomalies

Along with temporal changes, the spatial distribution of the gravitational field during periods of seismic activity plays an important role. The main focus in this regard was on the study of the most powerful earthquakes gravitational effects that were observed during the GRACE mission. One illustrative example is the time interval from October 2006 to January 2007, during which 7 earthquakes of M>5.0 were registered (Fig. 4). Over several months, starting in September 2006, slow changes in EWH (GFZ-RL06) gradually focused near the epicenter of the upcoming strong seismic events: November 8, 2006 (M=5.8), December 25, 2006 (M=6.2) and January 8, 2007 (M=5.8) (Fig. 2c).

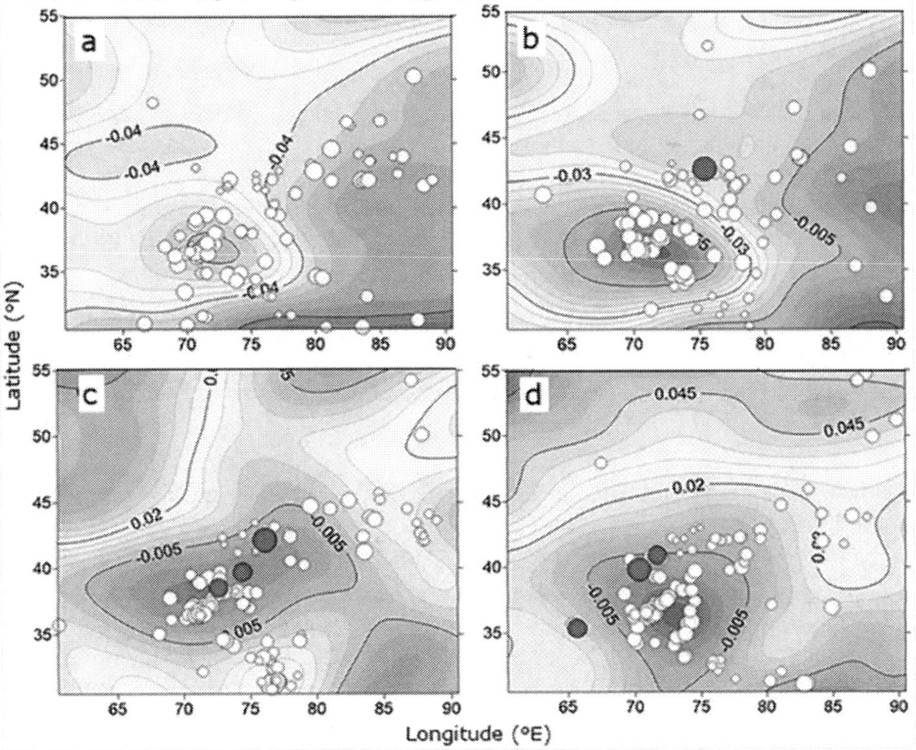


Fig. 4. Evolution of the gravitational field spatial distribution and seismicity in October (a), November (b), December (c) 2006 and January 2007 (d)

Gravitation field anomalies of seismic origin are usually associated with the displacement of the density boundaries and the effect of changes in density as a result of deformation (dilation, changes in the porosity structure) [10]. It can be assumed that the negative anomaly of the gravitational field observed in the junction zone of the Pamir and Tien-Shan, was associated with changes in the stress-strain state of the entire seismogenic region. Comparison of the gravitational field distribution with the location of M≥3.0

earthquake epicenters allows to conclude that the overwhelming number of events that occurred during the month spatially coincided with the anomalous zone EWH, which is characterized by a deformation process that increases the porosity (decompression) of the geological environment.

Similar results were obtained when constructing difference maps of the spatial distribution of gravitational field anomalies, at each point (pixel) of which the gravitational field value of the month of previous year was subtracted from the EWH values of the current month. Epicenters of major events also coincided with anomalous areas of decreased EWH values.

## 4. Conclusion

In this study, based on retrospective data from GRACE satellite measurements (2003-2015), we analyzed the dynamics of long-term changes in the gravitational field, presented in terms of the equivalent water level average monthly values above the geoid contour, and the relationship of these variations with the seismic process in the Pamir and Tien Shan territories. The research results allow to draw the following conclusions:

1. Studies of the earthquakes gravitational effects based on the use of various GRACE models allow to identify the gravity field component (ΔEWH), which may be associated with seismic activity.

2. The correlation between changes in the gravitational field and the geodynamic process as well as the probable seismic origin of positive / negative ΔEWH anomalies due to the redistribution of masses in the lithosphere during the preparation and passage of large regional ($M \geq 5.5$) earthquakes is established.

3. The spatial distribution of seismogenic zones coincides with the regions of negative anomalies of the gravitational field, which can be explained by the dynamics of lithosphere decompression.

4. The consistent behavior of long-term time series of equivalent water level (EWH) and the slope of the earthquake recurrence graph (*b-value*) reflects the intensity of geodynamic processes in a seismically active region.

5. The threshold for detecting gravitational effects of earthquakes is probably lower than the generally accepted level of magnitudes ($M>7.5$).

The study is partially performed within the framework of the state task of the Federal State Budgetary Institution of Science, Research Station of Russian Academy of Sciences in Bishkek (theme No. AAAA-A19-119020190064-9).

## References

1. S.I. Sherman, Geodynamics & Tectonophysics **7**, 495 (2016).
2. X. Qiao, Q. Wang, S. Yang, J. Li, R. Zou, K.Ding, Geodesy and Geodynamics **6**, 91 (2015).
3. U. Meyer, A. Jaggi, Y. Jean, G. Beutler, Geophys. J. Int. **2**, 1196 (2016).
4. J. Li, J. Chen, Z. Zhang, Earthquake Science **27**, 229 (2014).
5. URL:http://icgem.gfz-potsdam.de/
6. URL to Reproduce Results: http://www.thegraceplotter.com/
7. URL: http://www.isc.ac.uk
8. C. Narteau, S. Byrdina, P. Shebalin, D. Schorlemmer, Nature **462**, 642 (2009).
9. L.G. Sverdlik, Bulletin of the Kyrgyz-Russian Slavic University **18**, 187 (2018).
10. I. Panet, V. Mikhailov, M. Diament, F. Pollitz, G. King, O. de Viron, M. Holschneider, R. Biancale, J.-M. Lemoine, Geophys. J. Int. **171**, 177 (2007).

# Atmospheric temperature coherent variations effects, preceding strong earthquakes

*Leonid* Sverdlik[1,*], *Sanjar* Imashev[1]

[1]Research Station of the Russian Academy of Sciences in Bishkek, Kyrgyzstan

**Abstract.** We analyzed satellite measurements of temperature at upper troposphere/lower stratosphere (UTLS) levels separated by the tropopause above the epicentral area of a strong earthquake with a magnitude of M=7.3 in the Northern Tien-Shan (Kyrgyzstan) that took place on August 19, 1992. The developed algorithm and method of continuous wavelet transform allowed detecting of abnormal behavior, temporal, spatial and spectral coherence of short-period temperature variations, preceding the seismic event. The results show that the spatial structure and dynamics of temperature anomalies in the area of UTLS have a sufficiently stable relation to seismic activity.

## 1 Introduction

Progress in development and improvement of satellite technologies, as well as availability of a large number of specialized services and databases predetermined extensive use remote sensing results for purposes of research of the interaction between various geophysical processes and seismic monitoring in particular. An important task of anomalies identification in temperature time series (spatial images), together with selection of parameters, that may serve as potential attributes of strong earthquakes preparation, is selection and application of the most effective methods of processing of experimental data. A lot of methods and algorithms for selection of anomalies for various data types that describe processes occurring in time and preceding earthquakes, from simple methods of visual interpretation to more complex methods based on wavelet transform and conjoint application of wavelets and neural networks [1] have been developed. The most widely spread method of detection and localization of deviation of corresponding parameters from typical behavior is the RST (Robust Satellite Techniques) data analysis [2, 3].

Within researches of atmospheric effects conditioned by strong earthquake impacts an important physical characteristic, that describes dynamics and state of the atmosphere, is temperature. The task of selection of pre-seismic attributes of anomalous behavior in temperature time series may be solved on the basis of application of relevant algorithms [3–5] and the wavelet-analysis methodology with account for differences in frequency properties of spectra of anomalous temperature perturbations at upper troposphere/lower stratosphere levels separated by the tropopause (UTLS). The algorithm developed by the authors and

* Corresponding author: l.sverdlik@mail.ru

based on general principles of the universal RST method in combination with spectral and correlation analysis is intended for selection of short-period spectral constituents with consequent analysis of not only abnormality degree of amplitude but specifics of change in the phase and temporary realm, which are simultaneously observed at two UTLS levels divided by the tropopause. The output parameter of the algorithm, changes of which were compared to seismic activity, was integrated index of anomalous variations ($D_{\delta T}$), that actually served as the measure of coherence or conformity of temperature variations at two isobaric level in time and space.

In this research we present results of revealing of temporal, spatial and spectral coherence in UTLS temperature variations over the epicentral area of a strong earthquake of magnitude of M=7.3 that occurred in Kyrgyzstan on August 19, 1992.

## 2 Seismic and temperature measurements data

On August 19, 1992 at 02:04:37 (UTC) a strong earthquake with a magnitude of M=7.3 occurred at the depth of ~27 km in the Northern Tien-Shan (42.142°N; 73.575°E) when intense activation of the seismic process started [6]. A map of the aftershock field of the Suusamyr earthquake is presented in Fig. 1a. 20 aftershock of magnitude 3.8–6.3 were registered during the day. Fig. 1b shows a histogram of distribution of the total daily energy (Es) of regional seismic events registered in Kyrgyz Seismic Network (KNET) seismological network in August 1992. Sharp activation of seismic intensity was decreasing to background values at the end of 1993.

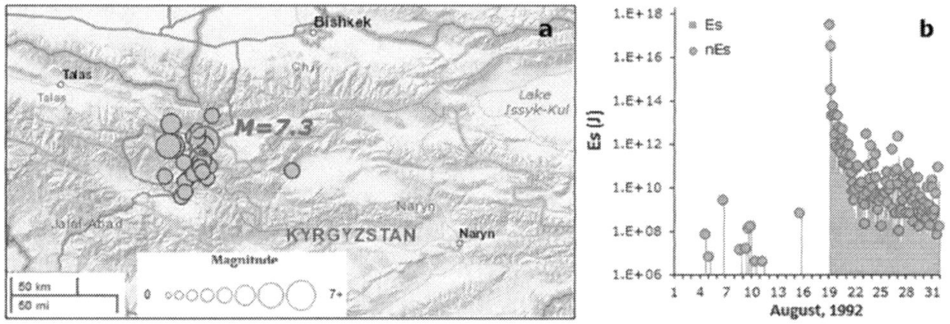


Fig. 1. The aftershock field of the earthquake of M=7.3 (Northern Tien-Shan) 19.08.1992 (a) and distribution of the total daily energy (Es) of regional seismic events in August 1992, (b)

Analysis of altitude and temporal variations of atmospheric temperature was performed using satellite measurement data. Temperature data of the MERRA-2 [7] reanalysis system are based on satellite observations processed in a global model and represent synthesized temperature values at standard isobaric levels from 450 to 70 hPa with spatial resolution of 0.5°×0.625°. An advantage of reanalysis is higher time discretization, that is Δt=3h, which allowed to detail the process of temperature perturbations formation.

## 3 Results and discussions

### 3.1 Temporal and spatial temperature perturbations

Given that the earthquake under consideration belongs to special events in the regional seismicity and is the strongest one (M=7.3) in the territory of Tien-Shan over the last 30 years, the whole annual cycle (1992) of temperature variations in the tropopause has been analyzed. Changes in semidiurnal temperature variations at isobaric levels from 300 hPa to

100 hPa in 1992 (Fig. 2a) were of complex nature and presented a seasonal trend modulated by various short-period components. Diagnostics results of temperature anomalies at the levels separated by the 300 hPa and 100 hPa tropopause using the algorithm described in [3, 5] demonstrated a well-defined values outburst of $D_{\delta T}$ = 2.5 parameter several days before the seismic event of M=7.3 (Fig. 2b), that allows to assume some relation with the earthquake preparation period (Fig. 2c).

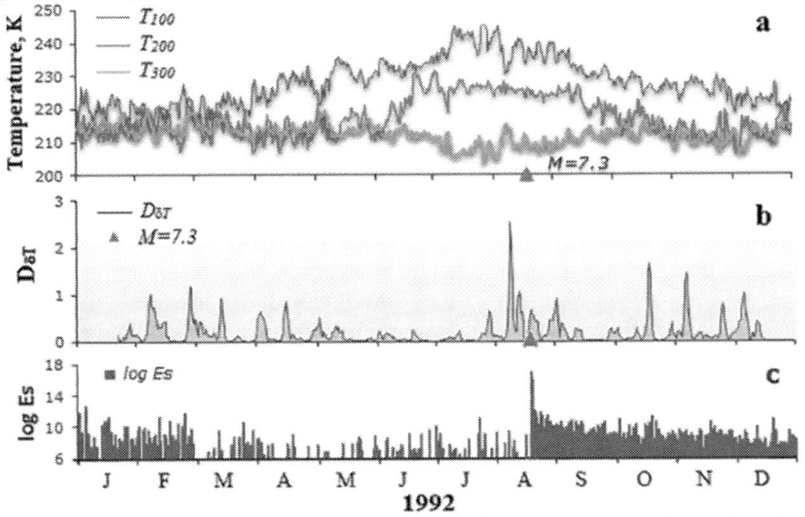


Fig. 2. Temperature variations at the 300, 200 and 100 hPa isobaric levels (a), changes of anomalous variations integral parameter $D_{\delta T}$ (b) and distribution of the log of radiated seismic energy (log Es) (c) in 1992

Changes of spatial distribution of $D_{\delta T}$ integral parameter, describing of the process of perturbations formation in the tropopause during preparation and occurrence of the strong earthquake on December 18–19, 1992 are presented in Fig. 3.

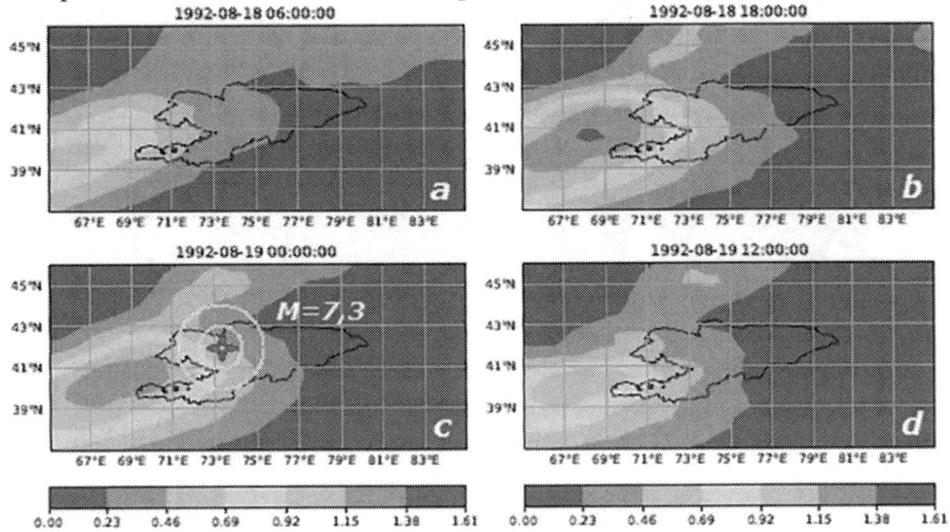


Fig. 3. Evolution of $D_{\delta T}$ integral parameter spatial distribution during the preparation and occurrence of the earthquake of M=7.3 (August 19, 1992, 02:04:36). The maximum of anomalous perturbation was observed ~8 h before the event

Location of the high values of $D_{\delta T}$ parameter remained unchanged for 2 days and the anomaly maximum was observed ~8 hours before the seismic event of M=7.3 (Fig. 3b).

As for the mechanism of formation of anomalous temperature perturbations preceding major earthquakes, we hold to an assumption that wave processes, i.e. effects of excitement of intense gravitational waves, in particular, long-period seismic-gravitational oscillations, play the main role in lithosphere-atmosphere interaction. Increase in intensity of oscillation of large local volumes of the Earth's crust in the process of development of strong earthquakes results in synchronous and spectrum-identical atmospheric pulsations, the recurrence rate of which is ~6–8 days [8].

An illustration of changes of altitude distribution and characteristic wavelengths of temperature perturbations is transformation of wave manifestations in vertical profiles (Fig. 4a, b) and respective temperature anomalies in wavelet spectra before and after the earthquake (August 18 and 21, 1992, 03:00 UTC). The altitude range from 5.0 km to 60.0–65.0 km was considered. The most intense harmonics preceded the seismic event and were observed in the UTLS. The main contribution to temperature variations was made by oscillations with vertical wavelength $\lambda_Z$=11.0–12.0 km (Fig. 4c). After the earthquake, temperature vertical distribution represented a more complex combination of two wavelengths (~6.0 km and 18.0 km). The most intense wave perturbations ($\lambda_Z$=15.0–20.0 km) moved to the stratopause (Fig. 4d).

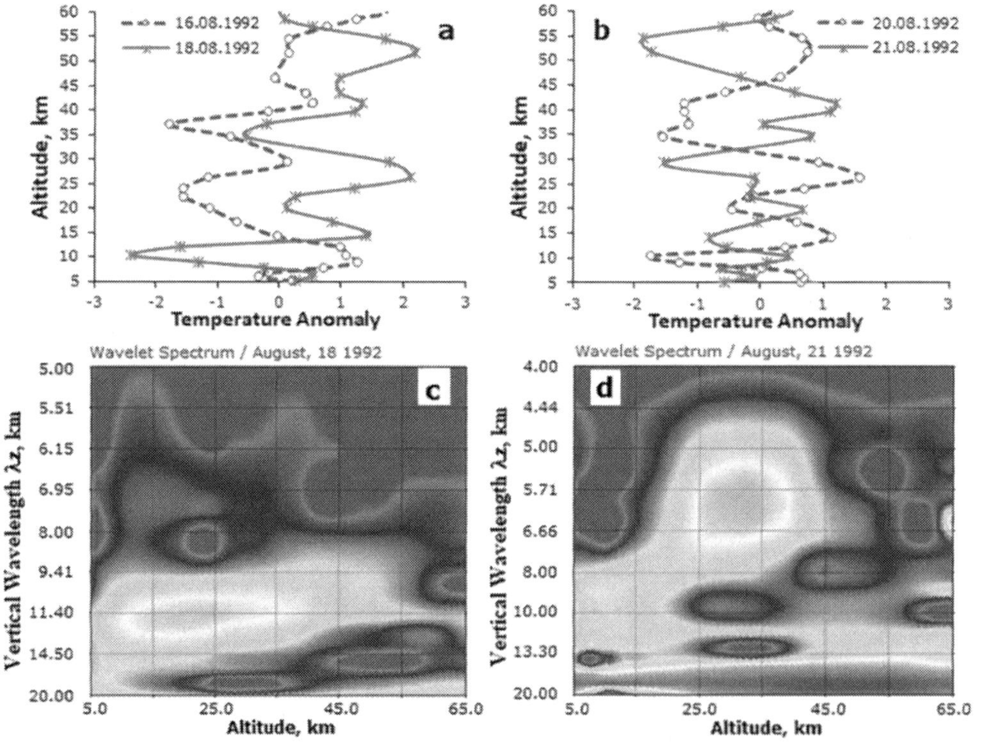


Fig. 4. Transformation of vertical temperature profiles and temperature anomalies wavelet spectra before and after the earthquake (M=7.3): (a) and (c) – August 18, 1992; (b) and (d) – August 21, 1992

### 3.2 Spectral coherence estimates of temperature variations

Analysis of dynamics of periodicities of satellite data analyzed was performed using continuous wavelet transform. Wavelet spectrograms of temperature time series at 300 hPa and 100 hPa levels (Fig. 5), built for the range of periods from 0.125 to 64 days, demonstrate change in frequency properties during the whole annual cycle under consideration. Specifics of the presented spectrograms are well-defined intensive oscillations within the range of periods of 4–12 days in first decades of August. Whereas, there wasn't any simultaneously appearing at both isobaric levels and equally significant temperature perturbation in the dynamics of variation spectrum in 1992.

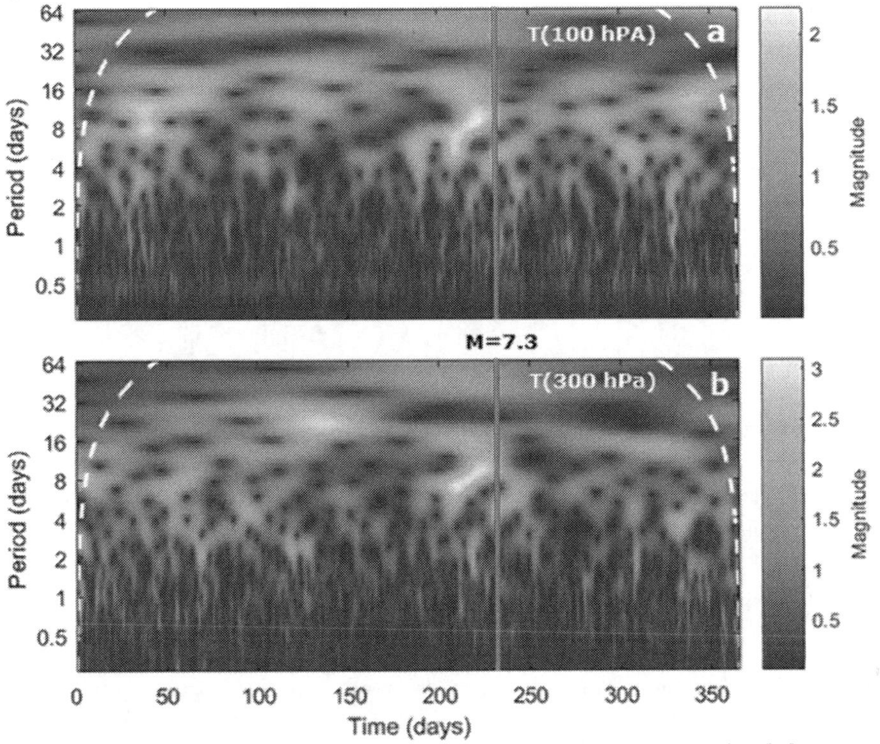


Fig. 5. Dynamics of temperature variations spectra at the 300 and 100 hPa levels in 1992

To analyze consistent variability and statistic interrelation between temperature variations in atmosphere layers divided by the tropopause above the epicentral area of an earthquake calculation of the spectral measure of temperature was performed (Fig. 6).

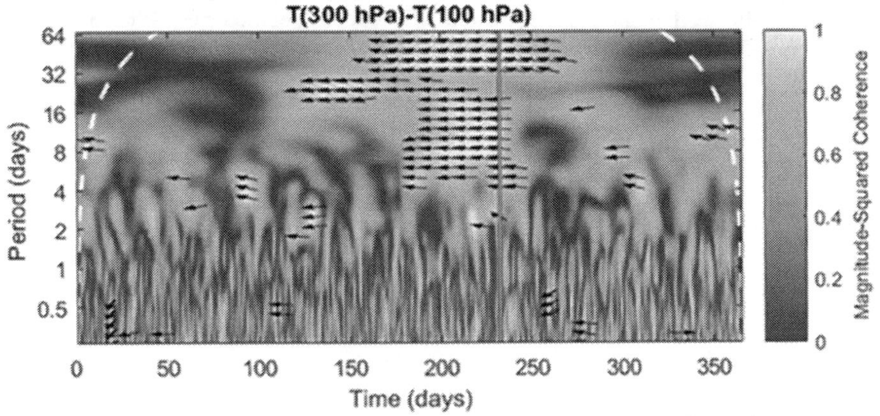


Fig. 6. The square of coherence between coefficients of wavelet transform of temperature of two time series at isobaric levels of 300 and 100 hPa in 1992. The vertical line is the moment of the earthquake of M=7.3

Temporal changes of coherence between coefficients of wavelet transform of temperature at isobaric levels of 300 hPa and 100 hPa presented in the diagram characterize correlations, that exist between individual spectral components.

The effect of strong coherence of temperature at the atmosphere levels under consideration preceded the strong earthquake and became apparent in the range of periods of 4–12 days. Thereat, the seismic-atmospheric effect of a strong earthquake was more distinctly apparent in high-frequency components of the spectrum, the period of which was from 4 to 6 days. Duration of anomalous increase of coherence was substantially longer for levels located close to the tropopause. Arrows directions are indicative of phase opposition of oscillations under observation. Therefore, occurrence of anomalous temperature perturbations is in line with the time of coherent behavior manifestation, and the highlighted range of periods, where the effect is most distinctly observed, coincides with experimentally obtained assessments of most informative spectral components, that we used in calculations of integral indicators [4].

## 4 Conclusion

Results of analysis of satellite remote sensing data showed that anomalous temperature changes in the tropopause were observed above the epicentral area of the strong earthquake of M=7.3. Study of periodicities dynamics revealed effects of coherent temperature behavior in the UTLS in the interval of periods from 4 to 6 days preceding activation of the seismic regime in the territory of the Northern Tien Shan. The presented spatial-temporal distributions of anomalous temperature perturbations point at probable relation to the seismic process. We observed differences in vertical distribution of temperature variations frequency characteristics before and after the earthquake.

On the basis of the results obtained it can be assumed that for the purpose of research of seismic-atmospheric effects of strong earthquakes, along with visualization of development of spatial distribution of the anomalous variations of integral parameter, generation of maps reflecting dynamics of temperature variations in short-period spectra, evolution of vertical wave lengths and density of potential energy of gravitational waves in the troposphere and stratosphere may be promising.

The author is grateful to Giovanni team of NASA GES DISC for free access to satellite measurement data and modeling results.

The study is partially performed within the framework of the state task of the Federal State Budgetary Institution of Science, Research Station of Russian Academy of Sciences in Bishkek (theme No. AAAA-A19-119020190064-9).

## References

1. Z-H. Jiao, J. Zhao, X. Shan, Nat. Hazards Earth Syst. Sci. **18**, 1013 (2018).
2. V. Tramutoli, G. Di Bello, N. Pergola, S. Piscitelli, Ann. Geophys-Italy. **44**, 295 (2001).
3. L.G. Sverdlik, S.A. Imashev, GEOSISTEMS of Transition Zones. **3**, 19 (2019).
4. L.G. Sverdlik, S.A. Imashev, J. Sib. Fed. Univ. Eng. Technol. **11**, 956 (2018).
5. L.G. Sverdlik, S.A. Imashev, MAUSAM. **71**, 481 (2020).
6. V. Kashkin, L. Sverdlik, R. Odintsov, T. Rubleva, K. Simonov, A. Romanov, S. Imashev, E3S Web of Conferences, **149**, 03011 (2020).
7. URL: https://disc.gsfc.nasa.gov/datasets/
8. E.M. Linkov, L.N. Petrova, K.Ts. Osipov, Doklady AN SSSR. **313**, 1095 (1990).

# Reflectometry of ice cover at different seasonal conditions by signals of navigation satellites in L1-range

*Anatoly* Sorokin[1]*, *Evgeniy* Kuzmin[2], *Daniil* Makarov[1]

[1]Federal Research Center KSC SB RAS, Krasnoyarsk, Russian Federation
[2]Siberian Federal University, Krasnoyarsk, Russian Federation

**Abstract.** A series recordings of interference reflectograms from a section of a lake with ice cover in the stage of freezing, stationary winter state, as well as during heating and melting was carried out. The presented experimental data were obtained in L1-range navigation satellites. The interference effects during periods of ice freezing, stationary state of ice (winter) and during melting (spring) were revealed. A variant of volumetric violation of the ice structure as a result of the influence of heat fluxes through the air-ice and ice-water interfaces is considered.

## 1 Introduction

The cycle of formation of ice covers of rivers and lakes in Siberia in natural conditions from freeze-up to the complete completion of the ice melting process lasts 5-6 months or more. The practical use of ice covers is quite developed: ice crossings, winter roads, ice on rivers, professional and amateur ice fishing. It is known that in spring the ice cover becomes less durable even with winter thickness, its crystal structure is saturated with water. Thus, the task of non-contact monitoring of the state of ice covers seems to be in demand and urgent. The solution to this problem is advisable using the available global navigation satellite systems (GNSS) GLONASS, GPS and GALILEO. The use of GNSS signals ensures the implementation of continuous monitoring of the earth's cover, atmosphere, and ionosphere. Methods of GNSS reflectometry with long-term registration of amplitude variations of signals from navigation satellites (NS) allows the restoration of the electro physical characteristics of natural environments [1, 2, 3].

## 2 The dynamics of ice cover conditions

River and lake ice are widely represented on planet Earth. A significant proportion of the areas of water bodies in Siberia in winter is covered with ice, its thickness varies over a wide range. The dynamics of freeze-up on rivers and lakes differs significantly under the same weather conditions. The determining factor for the growth rate of ice thickness is the sum of daily average negative temperatures [4]. The most homogeneous ice in terms of geometrical and electro physical characteristics is formed on small lake water bodies. Under conditions of stable cooling during the freeze-up period, smooth inner and outer ice

*Corresponding author: sorav@iph.krasn.ru

surfaces with a small number of inclusions of gas bubbles and other volume defects are formed. The option of using such objects for studying the processes of interaction of NS signals with ice covers in order to develop a method for diagnosing the mechanical strength of ice is convenient for the practice of relevant experiments.

## 3 Dielectric model of ice

The main radio physical characteristics of fresh water and ice differ. The complex dielectric constant (here in after - CDP) and the refractive index in the range L1 at temperatures of $0^oC$ have the following values: dry ice $\varepsilon' \sim 3{,}0\text{-}3{,}2$ and $\varepsilon'' \sim 0.0003$, water $\varepsilon' \sim 86$ and $\varepsilon'' \sim 7$. These characteristics of wet ice as a mixed dielectric made of ice grains and liquid water at temperatures near $0^oC$ interact weakly, which makes it possible to calculate the CDP of the wet ice layer. CDP of wet ice $\varepsilon'_{wi}$ is determined by the expression:

$$(\varepsilon'_{wi})^{\gamma} = W_i(\varepsilon'_i)^{\gamma} + W_w(\varepsilon'_w)^{\gamma}, \quad (1)$$

where $W_i$, $W_w$ are the volume fractions of ice and water in the water + ice mixture, $\gamma$ is the coefficient, varies in the interval $0 < \gamma \leq 1$, is determined experimentally or from theoretical premises.
The real and imaginary parts of the complex values $\varepsilon$ and n are determined by the relations:

$$\varepsilon' = \alpha^2 - \beta^2,\ \varepsilon'' = 2\alpha\beta. \quad (2)$$

Refractive and absorption indices are respectively determined by expressions:

$$\alpha = \sqrt{(\sqrt{(\varepsilon'^2 + \varepsilon''2)} + \varepsilon')} / \sqrt{2} \text{ and } \beta = \sqrt{(\sqrt{(\varepsilon'^2 + \varepsilon''^2)}} / \sqrt{2} \quad (3)$$

Characteristics of interference reflectograms of ice covers are determined by significant differences between $\alpha$ and $\beta$ of water and ice.
During the life cycle, the temperature field in the volume of the ice cover is dynamic. Temperature gradients at the ice - air - and ice - water boundaries in autumn and spring differ significantly. The heat flux during the freeze-up period from the lower boundary is directed vertically upward and ensures the growth of ice from below. The temperature regime ensures the translation of the 0 ° C temperature point in the ice volume together with the lower boundary during the period of increasing ice cover thickness. During this period, under the condition of a stable decrease in the average air temperature, the ice has a crystalline structure. An increase in air temperature and radiation heating of the upper ice surface in spring smoothly change the direction of the heat flow to the opposite. Localized micro-volumes of liquid water, partially associated with the ice surface, appear in the ice structure. The dielectric constant of surface-bound water decreases markedly in comparison with liquid water. [5,6] In the mixed dielectric refractive model, it is possible to consider wet ice as a three-component medium with a variable concentration of bound water.

## 4 Measurements and ice covers

The registration of interference reflectograms (IR) was carried out according to the traditional scheme [3] based on two types special of GNSS equipment. A set of antennas with linear and right-hand circular polarizations was used. The IR registration scheme is shown in Fig. 1. On the scale of the lake size, the wave front of NS signals is practically

flat. The direction of its propagation is indicated by the wave vector. An interference field is formed at the point where the antenna is located as a result of the addition of a direct wave and many reflected from the interfaces: air-ice, single reflection and ice-water, with a double passage through the ice mass and the air-ice interface.
The fig.1 shows the most intense wave. From subsequent reflected waves with a path difference of a shorter coherence length of the NS radiation, the signal powers decrease exponentially, their contribution to the formation of IR is small. At the point where the antenna is located, an interference field is formed as a result of the addition of the direct wave and many reflected ones.

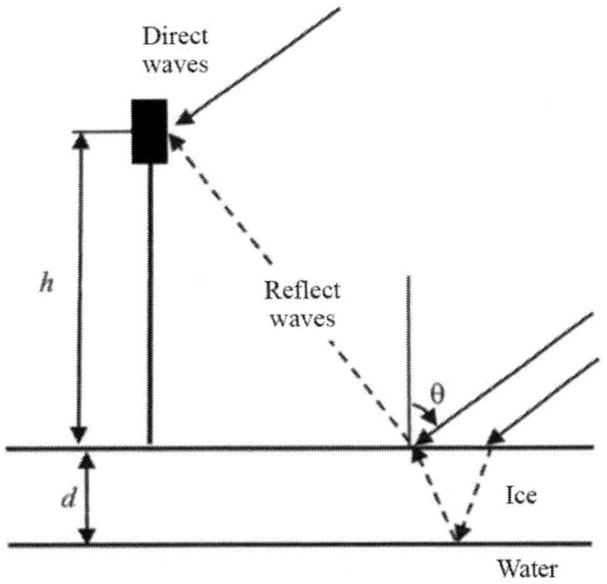


**Fig: 1.** Scheme of the formation of an interference signal at the receiving antenna

The powers of the signals reflected many times inside the ice significantly decrease; therefore, their contribution to the formation of interference reflectogram is small. Stable interference reflectograms of ice covers were recorded during the period from freeze-up to melting. The trajectories of the navigation satellites and the coordinates of the antenna location make it possible to unambiguously determine the coordinates of the "scanning" center for area reflection on the ice surface. Long-term registration (about 3-4 hour) of the interference reflectogram allows obtaining average data on the local ice thickness
Long-term registration (about 3-4 hour) of the interference reflectogram allows obtaining average data on the local ice thickness [1]. The strength qualities of "spring" ice, besides thickness, are determined by structural changes associated with the appearance of melt water in its volume. The ratio of the real part of the dielectric constants of water and ice near $0^{\circ}$ C in the L1-range is 26.9.

Seasonal changes in the strength characteristics of ice sheets are caused by the water – ice structural phase transitions. The formation of surface ice occurs rather quickly when the surface water layer is cooled to $0^{o}$C and there are no waves. Photographs of dry and wet ice samples are shown in Fig. 1, a) and 1, b), respectively.

***a)*** ***b)***

**Fig. 2**. Photos of ice samples a) newly formed ice (03.11.2018), b) wet ice (31.03.2019)

The visually observed ice cover was a transparent ice layer over the entire surface of the lake with rare extended continuous cracks. According to the classification of lake ice [4], it is lake-water ice, transparent, with air inclusions. In the static state of supercooled water, crystalline needles of ice appear on the surface - the initial forms of crystallization. As the air temperature decreases, the water at the ice-water interface cools and crystallization continues. The ice thickness increases, the initial process of ice needle formation is preserved and fixed in the ice mass. The appearance of local water inclusions during heating is most likely initiated by defects in the crystal structure of polycrystalline ice.

## 4 Results and discussion

A series of IR measurements using signals from the NS GPS and GLONASS in the L1-band (1.5-1.6 GHz) was carried out on a lake near Krasnoyarsk with an ice cover in the freeze-up stage (October-November), and in the spring, during the warming up and melting (March-April). A interval of ice thickness varied from 4.2 cm to 105 cm. Receiving antennas were installed on the ice, or on the coast (during the melting period) at the ice edge. The height of the location of the antennas varied in the interval 3.8 m - 4.4 m.
In fig. 2 show the IR where φ is the elevation angle for ice thicknesses is 4.2 cm (03.11.2018) and 105 cm (31.03.2019), respectively.

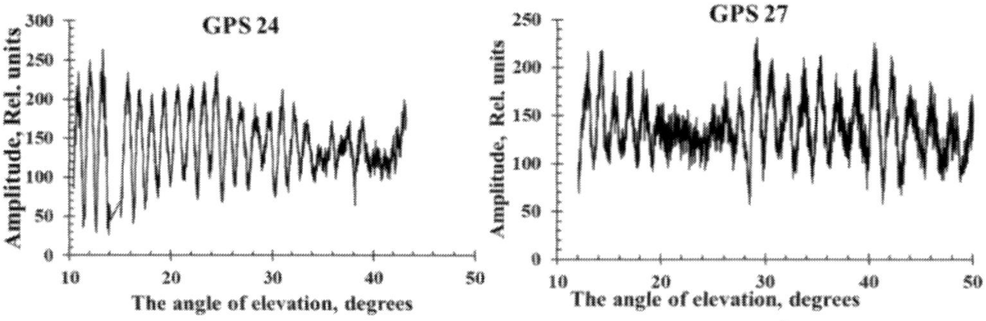


*a* *b*

**Fig. 2.** Experimental IR: a) newly formed ice (03.11.2018); b) wet ice, (31.03.2019)

In Fig. 3 show the experimental IR where ice thicknesses of 64 cm (02.04.2017) and 80 cm (20.04.2019), respectively

https://doi.org/10.1051/e3sconf/202022303016

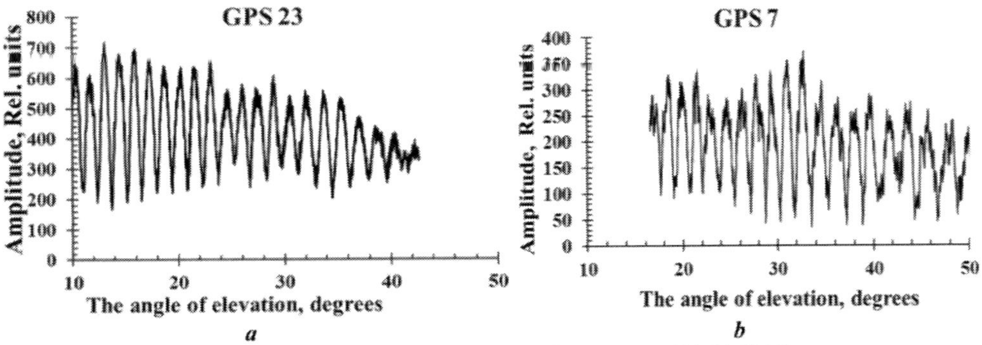


**Fig. 3.** Experimental IR: a) dry ice, (02.04.2017); b) dry ice + water, (20.04.2019)

As seen from Fig. 2 and 3, experimental IRs have a visually distinguishable modulation structure and noticeable features variation of the amplitude. The parametres IR are determined by the ice thickness and its moisture content. The spatio-temporal the amplitude variations of the total signal are formated by the superposition of the direct and probe signals. Its possible because interaction radio wave with two borders air-ice and ice - water and propogates in ice volume. This process is defined by of signal and depends of the refractive, absorption, thickness and condition of ice surfaces, at the amount and distribution of melt water in its volume.

A model of wet ice as a mixed dielectric can be a three-component medium with a variable concentration of two and more types of water capsules - bulk and surface [7,8]. The temperature field with a vertical gradient in the volume of lake ice forms layers of varying degrees of humidity and thickness. In the freezing stage at a stable negative average daily air temperature, a monolayer model of a rapidly growing dry ice layer is real. At the ice-water boundary, a thin layer of wet ice may form, slowly growing with increasing hydrostatic pressure on the lower boundary of the ice cover. When the ice thickness stabilizes and the surface of the air-ice boundary begins to warm up, an upper layer of wet ice grows downwards. For some time, three layers co-exist in the ice thickness: the growing wet upper and lower layers, and the decreasing dry middle layer

In IR processing, it is necessary to take into account the influence of coherence length of proube signals. The probing signals of the NS GPS and GLONASS in the frequency range 1.5 .-. 1.6 GHz), have similar values of the signal power in the near-Earth space, the type of modulation, and the signal repetition period. The differences lie in the method of separating the signals of individual satellites and the width (signal frequency bands: ~ 2 MHz for GPS and ~ 1 MHz for GLONASS). Accordingly, the coherence lengths are: GPS ~ 300 m, GLONASS ~ 600 m [1, 2].

## 5 Conclusions

A principal possibility of the method for monitoring the state of ice covers is due to a significant difference in the complex permittivity of water and ice. The presence of melt water in the ice structure determines the refractive characteristics of ice and the parameters of experimental interference reflectograms. The process of ice formation and melting is determined by vertical heat flows through the air-ice and ice-water boundaries. A layered refractive model adapted to the ice life cycle is promising in data processing algorithms.

The use of GNSS specialized equipment provides registration of amplitude variations of navigation satellite signals for several hours and obtaining the corresponding array of data on the state of ice.

## References

1. S. Jin, E., Cardellach F.Xie *GNSS Remote Sensing.* (New York, London: Springer Dordrecht Heidelberg, 2014)
2. D. S.Makarov, D. V.Kharlamov A. V. Sorokin *Siberian Journal of Science and Technology*. Vol. **20,** 1 (2019), Doi: 10.31772/2587-6066-2019-20-1-8-19
3. M.I. Mikhaylovm, K.V. Muzalevsry Mjronov V.L, *Modern problems of remote sensing of the Earth from space.* **14,** 2, pp 167-174.
4. V.V. Bogorodsky, V.P.Gavrilo, *Ice. Physical properties. Modern methods of glaciology.* (Gidrometeoizdat, Leningrad, 1980)
5. Yu.I. Lukin, V. L. Mironov, S. A. Komarov, Izvestiya vuzov. Physics, 51, 9 (2008)
6. V.L. Mironov, Yu.I. Lukin, Izvestiya vuzov. Physics, **53**, 9 (2010)
7. G.S. Bordonsky, A.O.Orlov, Letters to ZhTF, **45, 5** (2019)
8. Nilsson, L.G.M. Pettersson, Chemical Physics 389 (2011)

# Peculiarities the navigation satellites signals scattering in the layered structure of the tree stand

*Vladimir* Podoprigora[1,3,*], *Anatoly* Sorokin[2], *Dmitriy* Kharlamov[2]

[1] L.V.Kirensky Institute of Physics SB RAS, Krasnoyarsk, Russian Federation

[2] Federal Research Center KSC SB RAS, Krasnoyarsk, Russian Federation

[3] Siberian Federal University, Krasnoyarsk, Krasnoyarsk, Russian Federation

**Abstract.** A scattering and attenuation processes of navigation satellite signals in the forest interaction with tree trank and crown layers a tree stand are presented. The influence of the boundaries soil - tree stand – crown - air on the characteristics of radio signals scattering after interaction with forest is considered. Structural and electro physical features of tree stand elements in the processes of interaction with signals of the L 1 range and experimental data on signal transformations the forest are analyzed.

## 1 Introduction

In recent years, for the purpose of remote sensing of the earth's covers, radio wave measurements have been successfully used, effectively using the capabilities of the global navigation satellite system. To extract information about micro- and macro parameters of tree stands from satellite signals processed after passing through the tree stand. It is necessary to use a model of a electromagnetic field interaction with tree elements. It is known that at present there is no universal mechanism that takes into account: the multiscale of the forest environment, different radio frequency ranges, the configuration of the experiment, the spatial distribution of trees, the orientation of branches, wood moisture, dielectric properties of wood, bark, leaves and needles. This explains the large number of works that, to one degree or another, explain the features of the passage of a radio signal in a forest, but carried out at various objects in different conditions [1-2]. However, the interaction of radio emission with a tree stand is considered mainly for radar measurements at very low and very high frequencies and without taking into account the influence on the parameters of orientations of tree elements, anisotropy and interfaces of different media.

The peculiarity of the satellite radiation we use is that the wavelength in the L1-range of radio frequencies (1.5–1.6 Hz, $\lambda \approx 19$ cm) is much larger than the transverse dimensions of the needles and leaves, but is comparable to the diameters of the trunks and thick branches of a tree. In this frequency range, the peculiarities of the internal structure of trees are revealed, which makes it possible to recognize their species and types of tree crowns

* Corresponding author: podoprigora46@gmail.com

using remote sensing. The maximum effective scattering area coefficient for conifers is also observed at these frequencies.

The purpose of this work is to study the parameters of the scattering of signals from L1-band navigation satellites from the selected layers of tree trunks and crowns, taking into account the influence on these parameters of the orientation of tree elements and the boundary "soil - layers of trunks and crowns.

The scattering and attenuation characteristics were investigated for forward scattering of radio signals from GLONASS and GPS navigation satellites at a frequency of 1.5-1.6 GHz. The description of the measuring equipment and the scheme of the experimental study are given earlier [3].

## 2 Model of interaction the navigation satellites signals with the tree stand layers.

The tree elements were represented as discrete scatterers, which, in contrast to the continuous medium model, makes it possible to detect spatial inhomogeneities (density fluctuations) in the tree stand. A model description of the interaction of NS signals with a forest assumes the selection of areas with a certain ratio of the sizes of tree elements to the wavelength of radio waves. This allows the correct use of the corresponding approximation. Mechanisms of scattering and attenuation of radio waves in the upper and lower parts of the tree are significantly different. It is natural to select as the above areas the layer of trunks and the layer of crowns, consisting of branches and needles or leaves.

Each layer of the tree stand differs from the other in density, shape, size of tree elements and their orientation in space. The trunks layer is considered as infinite dielectric cylinders, large branches - as cylinders of finite length, needles - highly elongated ellipsoids. To describe small scatterers, the generalized Rayleigh - Hans - Debye (RGD) approximation for ellipsoids was used, which requires the relation $k_0 d(n-1) << 1$, where $k_0 = 2\pi/\lambda$ is the wave number in free space, $d$ is the smallest size of the ellipsoid, $n$ is the refractive index. The scattering medium is more transparent for radio waves at least in the direction for one of the ellipsoid axes.

Scattering on trunks and thick branches of trees corresponds to the approximation of geometric optics for infinite and finite cylinders [4] with the latter's size restrictions: $h>>r$, $k_0 h>>1$, where $h$ and $r$ are the height and radius of the cylinder. It was assumed that the internal field of a finite scatterer can be considered as a field inside an infinite cylinder. When solving the diffraction problem on a tree trunk, this approximation is fulfilled already at $h/r > 5$ [4].

An important role in the process of radio wave propagation in the forest is played by the orientation of the scatterers relative to the incident and scattered rays. It can be taken into account in the framework of the tiered approach [5]. The position of the crown branches in space is described by the tensor of orientational ordering, and the components of the scattering cross section tensor are proportional to the square of the anisotropy and the components of this tensor. Here, the tiers play the role of crown sublayers, in each of which there is a uniform distribution of thick branches along the azimuthal angle. The angle of their inclination to the vertical changes with a sinusoidal distribution function from 30° to 120° - from the upper to the lower tiers. A separate layer or tier of the crown consists of groups of scatterers, each of which is characterized by a certain size and orientation of thin branches with its own distribution function.

## 3 Boundaries of layers in a tree stand

At sufficiently high frequencies (above 1 GHz), the description model of the scattering and attenuation processes of radio waves in a forest can include irregularities of different scales of interfaces, such as "soil – tree trunks", "crown - air", "tree bark - air". These irregularities can affect the scattering characteristics when considering the expressions responsible for the interaction of waves with the indicated boundaries. The shape of the reflecting surface of these boundaries is always irregular, and the height of the points of the surface *h (x, y)* is described by a random function of coordinates (statistically rough surface). It is usually considered that the distribution of heights h obeys the normal law with the mathematical expectation $\langle h(x,y)\rangle = 0$ and the variance $\delta_h^2$. The spatial correlation function $L_h = \langle h(x_1,y_1)\times h(x_2,y_2)\rangle$ is assumed to be isotropic; a Gaussian curve is often used for it.

A surface is considered to be fine-rough if the height $h(x, y)/\lambda$ and its partial derivatives with respect to the $x$ and $y$ coordinates are much less than one. In some cases, such surfaces include the boundary "soil - layer of trunks". For the scattering matrix, the well-known Kirhoff model (with the constraint $h/\lambda<0{,}3$) can be applied.

Large-scale irregularities (for example, the boundary "air - crowns" for the $L1$-frequency) are characterized by the radii of curvature $R_{x,y} >> \lambda$, while no limitation is imposed on their heights. For large irregularities $\delta_h/\lambda = h/\lambda >> 1$, variance $h^2 \leq \lambda^2$ or $h^2 \geq \lambda^2$. With these assumptions the scattering characteristics can be calculated.

The study shows that if, when determining the components of the scattering matrix, the interactions of waves with elements of crowns, trunks and soil surface are taken into account. Soil irregularities can significantly change the angular dependence of these components. So, modeling of the dependence of the specific effective scattering area (SESA) on the angle of incidence, taking into account the interaction of waves with the soil and the layers of the tree stand, shows the same values of □$_0$ for a smooth and rough surface in the range of incidence angles □ = 5□-7□. In the region □ = 30° – 40°, the difference in the SESA values reaches almost 9 dB, the contributions to scattering from interaction with the soil are higher for an uneven surface than for a flat one, and strongly depend on the size of roughness and the polarization of the incident radiation.

## 4 Radioscopy the volume of larch forest by signals a navigation satellites

Registration of a scattered and attenuation signal by a larch stand (the Pogorelsky Bor station, Institute of Forest, KSC SB RAS) was carried out by a receiving antenna with a vertically up top oriented pattern. The antenna was located above the soil at a height of 4.2 m. The amplitude of the signal-to-noise ratio of right circular polarization and the angular coordinates of the satellites with a frequency of 1 Hz were recorded. The duration of the session was 23.6 hours. Inhomogeneities in the distribution of trees elements in the radio scanning zone of the tree stand transformed the parameters signals of navigation satellites.

The heterogeneity of the tree elements distribution in the area of radio scanning of the tree stand has the following characteristics: stand density - 0.114 trees / m2, average values of height and diameter are 26 and 0,19 meters respectively. The boundaries of the crown layer are located at heights from 12-13 to 25-26 meters and have vertical gradients biomass. The trunks layer occupies space from the soil to the gradient border with the height of the trank trees, at which the trunk diameter is comparable to the diameters of the main branches and is located at a height of 20-23 meters. The space at heights from 12 to 22 meters occupy parts layers crowns and tree trunks. The scanning signals of navigation satellites sequentially propogate through the both layers. The ratio of contributions to the change in signals parameters depends on the height of the receiving antennas and the elevation angle

The separation of the contributions to the field of scattered radiation by the crown layer and trunks is in principle possible with the use of coordinated electrophysical models of the layers and the technique of measuring the characteristics of the signals transmitted by the tree stand.

The graphs of space-time dependences of the signals characteristics of the GLONASS 9 navigation satellite after passing throuh thr forest shown in Fig. 1 and Fig. 2.

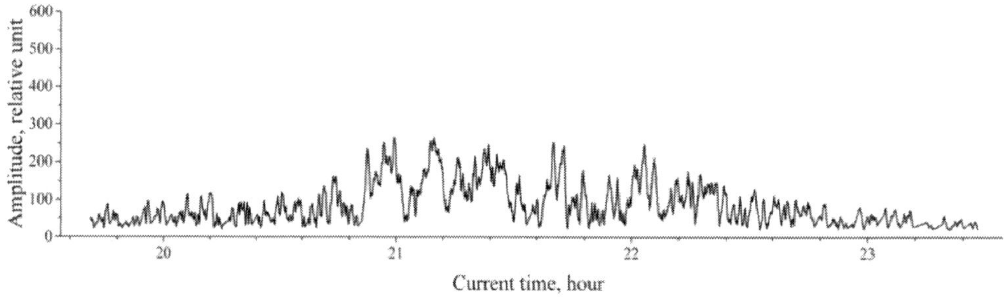


Fig 1. The space-time dependences amplitude of the signals after tree stand, session № 1

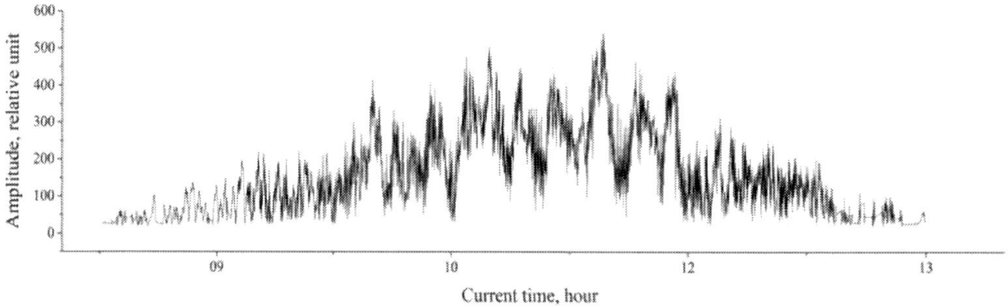


Fig 2. The space-time dependences amplitude of the signals after tree stand, session № 2

The tree stand was scanned in the sectors of azimuthal angles: session 1 (103°-19°) and session 2 (303° - 190°) with maximum elevation angles 31° and 75°, respectively. Option 1 demonstrates greater attenuation when driving through the forest compared to option 2. The Amplitude variations of the scanning signal are random and determined by the inhomogeneities of the tree stand.

The coherence of the probe radiation and the chaotically inhomogeneous distribution of the effective dielectric constant of the crown and trunk layers ensure the formation of a speckle structure in the radiation field scattered by the medium. A large amount of experimental data with a coordinate reference [5, 6, 10)] makes it possible to assess the state of local forest volumes.

In the process of scanning the forest stand with navigation satellite signals, it is effective to use different radio illumination options. These features are realized by moving the signal source and continuously registering the satellite's angular coordinates. The polar diagram of the angular coordinates of GLONASS 9 relative to the receiving antenna during radio illumination is shown in Fig. 3. The presented curves in figures 1 and 2combination with the coordinate data of the antenna location, the position of the radiation source and the

General taximetric characteristics of the tree stand allow you to determine the location of the scanning volumes and the linear attenuation coefficients in them.

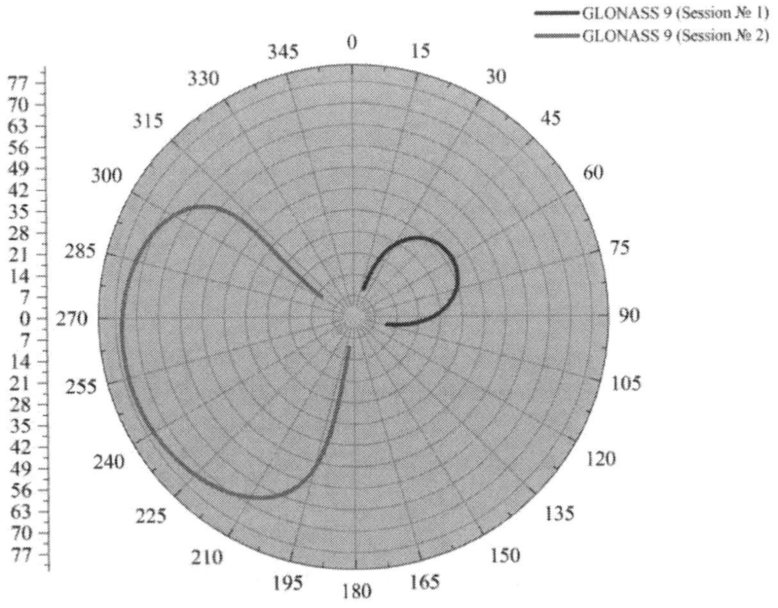


Fig. 3. The angular coordinates of GLONASS 9 in the process of radio transmission of the forest.

The complexity of modeling the propagation of radio waves in the forest is known [1, 2, 7, 8]. The prospects of using the signals of navigation satellites of the L1 range as a universal instrumental measuring resource and adaptive modeling of the forest stand, taking into account its layered structure, are real. An important factor in the implementation of the method for restoring the electro physical characteristics of the forest is to take into account the polystructure of the tree stand, [9], the anisotropy of the trunk layer [10] and the fractal structure of the crowns [11].

## Conclusions

Different-scale groups of scatterers in the crowns of trees with respect to the wavelength of a radio wave in the L 1 range (~19 cm) were identified: small branches, needles, leaves; large branches taking into account their spatial orientation.

The randomly inhomogeneous distribution of the effective dielectric constant of the crown and trunk layers in combination with the coherence of signals from navigation satellites provides the formation of the speckle structure of the radiation scattered by the forest.

An example of the space-time dependence of the field value at the point of location of the receiving antenna after the passage of the signal from the navigation satellite of the larch stand is presented.

A variant of restoring the electrophysical characteristics of the tree stand using signals from navigation satellites and modeling the process of propagation of radio waves, taking into account the polystructurality of the stand, anisotropy of the stem layer] and the fractal structure of crowns is real.

## References

1. V.I.. Popov, Propagation of radio waves in forests.-M. Hotline-Teletom. (2015).
2. A. Chukhlantsev. A.M Shutko., S.P Golovanov. Radio engineering and electronics.,V 48, No. 11, p.1285-1311 (2003)
3. V. Podoprigora, A. Sorokin. E3S Web Conf.,Volume 75, 01007, Regional Problems of Earth Remote Sensing (RPERS 2018) https://doi.org/10.1051/e3sconf/20197501007. (2019)
4. S. Seker, A. Schreider. IEEE Trasn.Antennas Propagat. 1988, V 36, p.303-307.
5. A. Sorokin, A Ostylovsky, A. Borisov, V. Ivanov, D. Makarov, D Kharlamov. E3S Web of Conferences 149, 0 0 https://doi.org/10.1051/e3sconf /20201490 RPERS 2019 20 6 2006 (2020)
6. V. Podoprigora, A. Sorokin, D. Makarov, D. Kharlamov. E3S Web of Conferences 149, 0 0 https://doi.org/10.1051/e3sconf /20201490 RPERS 2019 20 5 2005 (2020)
7. S. Novik, A. Zavyalov, E. Telpukhovsky, Izvestiya Vuzov. Physics. V.49, № 7, P.57-67. (2006)
8. A. Sorokin, V. Podoprigora, S. Fomin, I Savin. Izvestiya vuzov. Physics.. V. 60. № 12/2. pp. 19-22, (2017)
9. G. Karev, Structural models of forest ecosystems. Siberian Ecological Journal. 4 pp. 381-396. (1999)A.
10. Sorokin, V. Podoprigora,, D. S. Makarov, D.V. Kharlamov, V.V. Baltais. Proceedings of higher educational institutions. Physics. V. 63, No. 2. P. 50-54 (2020)
11. V. Yakubov, N. Lisenko. Izvestiya vuzov. Physics.. V. 55. № 12/2. pp. 19-22, pp 259-261. (2012)

E3S Web of Conferences **223**, 03018 (2020) https://doi.org/10.1051/e3sconf/202022303018
*RPERS 2020*

# Use of navigation satellite signals to determine stocks of stem wood

*Daniil* Makarov [1*], *Aleksandr* Borisov[2], *Anatoly* Ostylovsky[3], *Viktor* Ivanov[2], [1], *Dmitry* Kharlamov[1]

[1]Federal Research Center KSC SB RAS, Krasnoyarsk, Russian Federation
[2]Institute of Forest, KSC SB RAS, Krasnoyarsk, Russian Federation,
[3]Siberian Federal University, Russian Federation

**Abstract.** The method for determining the stocks of stem wood using the measurement data of the linear attenuation coefficients of signals navigation satellites L1 range and general taxometric characteristics of the forest is presented. A variant of determining the stocks of stem wood by differences in contributions to the attenuation by layers of crowns and trunks is considered, taking into account the coordinate referencing of the trajectory of the probe signal in the tree stand.

## 1 Introduction

The task of restoring stocks of stem wood by remote method using navigation satellite (NS) signals has the prospect of non-contact obtaining the average value of the cross-sectional area of trunks in the forest area. This indicator is determined based on the results of taxation using the regularities of the distribution of phytomass in the forest stand. The calculation of the sum of the cross-sections of the trunks is carried out according to the data of direct measurements of the diameters of the trunks at a height of 1.3 m, their distribution by diameters per 1 hectare of the forest stand. The structure of tree crowns and trunks is physically significantly different. The wavelength of radio emission of the NS L1 range is about 19 cm, which is much larger than the linear dimensions of the elements of the crowns - needles, leaves and small branches. The average sizes of the tree trunk diameters are in the range of 6 - 28 cm. The structure of the stand forest as a mixed dielectric made of air and trees is spatially inhomogeneous in the distribution of tree elements and trees themselves, has a layered structure [1-2]. Therefore, the contributions of scattering and absorption effects to the attenuation of NS signals are different. The crown layer is semitransparent, the signal attenuation is due to the absorption of the crown elements. The volume of layer, along with absorption, is an inhomogeneous scattering medium. The study of the possibilities of using radio scanning of the forest with NS signals is an urgent task.

## 2 Method of radio scanning the forest stand

* Corresponding author: mcdan3991@yandex.ru

Radioscopy of forest areas using NS signals is carried out by registering signals from navigation satellites after passing through the stand. The measurement scheme in the variant of the antenna location near the border of the forest [3] is shown in Fig. 1.

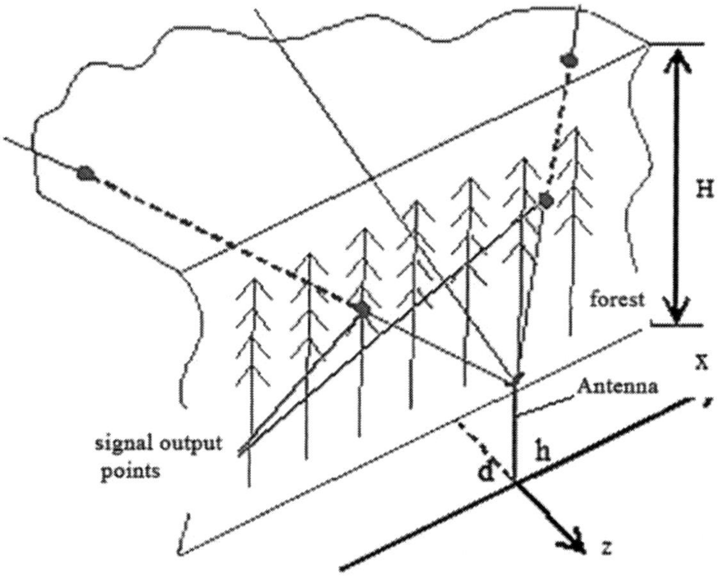


**Fig 1.** Scheme of the experiment on radio scanning of the tree stand.

Coordinate referencing of the signal trajectories that passed through the forest canopy makes it possible to unambiguously determine the spatial position of the analytical cone volume that forms the recorded signal. The path shown by the solid line corresponds to the registration of the direct signal that passed above the upper edge of the forest. Two paths pass through the stand in the crown layer and are indicated by dotted lines. With the change a height the antenna location h and its distance d to forest border. The contributions of crowns and trunks layers from scattering and absorption to the linear attenuation coefficients vary on different trajectories.

The method for determining the volumetric density of the forest stand biomass is based on the exponential law of attenuation of the signals of navigation satellites in the conical channel in the forest stand with the top on the receiving antenna and the solid angle corresponding to the first Fresnel zone. The antenna location in the forest stand with a vertically is oriented directional pattern axis. This allows to scan the space above the point of the antenna location in a conical volume with an angle 160°. Recording the probing radiation of individual satellites are realized in the analytical volume $V_F$. This volume is determined by the magnitude $(H-h)$. The total loss of signal power during propagation in the forest is determined by absorption and multiple scattering on trunks, branches, needles or leaves.

In [1], an expression was obtained for the share of the forest stand biomass $\mu$ in the analytical volume $V_F$, expressed in terms of the linear attenuation coefficient $\nu$ of the medium and the ratio of the power of the signal falling on the stand and leaving it. The location of the analytical volume is determined unambiguously from the coordinate data of the signal trajectory and is registered by the receiver. Changing the position of the antenna in the crown layer makes it possible to obtain the linear attenuation coefficients on the paths in the upper part of the crown layer relative to the antenna location. As the antenna height decreases, the contribution from the upper part of the stem wood layer appears in the coefficients. In general, the bulk density of stem wood is much higher and therefore the contribution to signal attenuation will be greatest when the antenna is located near the soil surface. Analysis of the dependence of the linear attenuation coefficient on the height $h$ and taking into account the contribution from the crown layer will make it possible to determine the linear attenuation coefficient in the stem wood layer.

## 3 An example of a test site

The selection of a test site is defined by minimum number of factors affecting the interection with the NS signals. Main biometric indicators of the forest stand with small dispresy from middle are very convenient. Tree stand with small values of deviations biometric indicators from mean values are the most optimal.

This criterion is met by a even-aged pine stand based (test site) in the Pogorelsky Bor, territorially part of the Krasnoyarsk sporadic forest-steppe. The pine forest at the age of 55 is represented by a highly productive tree stand of the 1st quality class. The spatial distribution of trees on the site is uniform, without glades and pronounced tree biogroups. Undergrowth, undergrowth and dwarf shrub layer are absent (Fig. 2).

**Fig 2.** Arrangement of trees on the test plot: the structure of the trunk layer.

The structure of tree crowns and trunks is physically significantly different, and the parameters of the stand that affect the signal attenuation will be different. It is necessary to have two main indicators - the basal area of the trunks and the phytomass of tree crowns in the upper part to obtain more objective characteristics of signal attenuation in a plantation.

The methodology performing scientific research was used to test site [4]. The characteristics of the trees in the test site were measured. They are: the average diameter at breast height is 18.6 cm; the average height of the stand is 20.7 m; the average diameter of tree crowns is 2.7 m; stand density - 1950 pcs / ha; the basal area - 53 $m^2$ / ha. The stock of stem wood on the RP calculated directly from the root area is - 514 $m^3$ / ha. Trees data were used to calculate for other characteristics. Crown sizes vary depending on the intensity of competition for available resources [5]. The result of such competition on the test site is shown in Fig. 3.

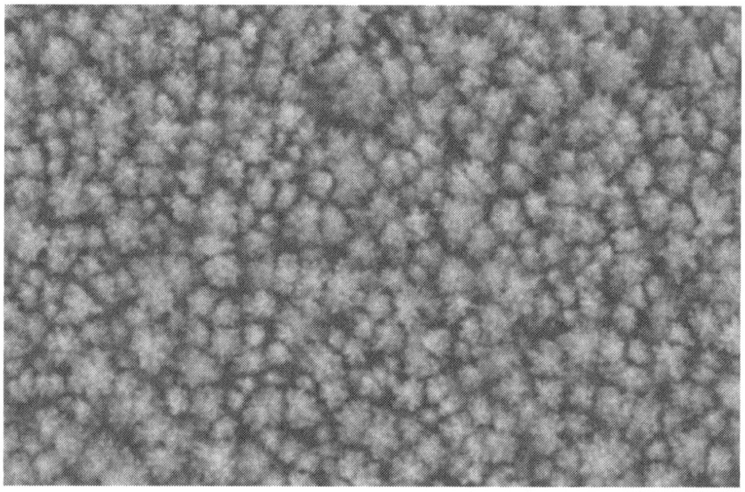

**Fig 3**. Arrangement of trees on the test plot: top view of the pine stand.

The size of trees by trunk diameter has significant differentiation for the same reason (Fig. 4)

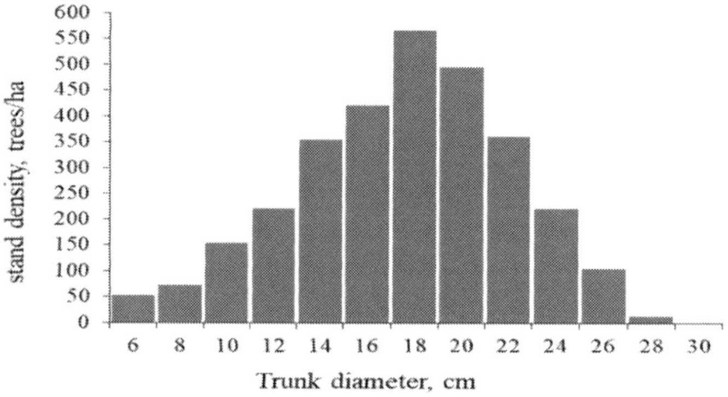


**Fig 4.** Distribution of trees by trunk diameter on the test site.

The ratio of crown and trunk biomass depends on the tree species, age, stand density, and available resource [6 - 8]. This ratio for sites based on Pogorelsklm Bor is about 1:6. By the height of the trees, taking into account the coefficient of the shape of the trunk, you can calculate the cross-sectional area of the trunk for a tree at any height. Taking into account the contributions of biomass components to the attenuation of the NS signal, depending on the registration angle and the antenna height, makes it possible to obtain the cross-sectional area of the trunks and the volume of stem wood in a plantation with an area of up to 3 hectares.

## 3 Conclusions

Instrumental methods for determining the stocks of stem wood using NS signals involve measuring the coefficients of linear attenuation in a representative array of the forest test site. Moving navigation satellites in the range of elevation angles from 10° to 90° and azimuth from 0° to 360° and the variable height of the antenna location allow to conduct a three-dimensional scan separately of the crown volume and the mixed volume of the crown and trunk layers.
The wavelength of the carrier radio wave in the L1 ~ 19 cm range practically coincides with the average diameter of the trunks and is noticeably larger than the characteristic dimensions of individual crown elements. This circumstance allows us to separate the contributions to the coefficients of linear attenuation in the layers of crowns and trunks. An additional factor in the processes of signal attenuation is the macroscopic heterogeneity of the tree density distribution in the forest area.
Radio probing by NS signals of the forest area provides registration of space-time coordinates of the NS position, recording time, amplitude of the signal-to-noise ratio with a frequency of 1 Hz. This information, combined with field measurements of the height of the forest, the coordinates of the location of the receiving antenna in the forest, make it possible to unambiguously determine the trajectory of the path and the linear attenuation coefficient of the NS signal in the forest.
The signal attenuation is due to the aggregate scattered on large-scale irregularities (tree trunks, large volumes of snow on the branches, irregularities in their distribution in space) and is partially absorbed in small-scale elements of trees (thin branches, needles, leaves) [9-11]. An additional effect on the recorded signal is made by the effect of depolarization of the GNSS signal and the appearance of a cross-polarized component due to the anisotropy of the permittivity of the forest.
The use of scanning the layered structure of the crown-stem tree stands allows one to separate their contributions to signal attenuation. The placement of the receiving antenna in the stand and taking into account the geometric characteristics of its layered structure allows one to obtain linear attenuation coefficients averaged over the volumes of crown and trunk layers. The array of obtained data has the prospect of solving the inverse problem of restoring a state of the tree stand, assessing the density of biomass and stock of stem wood.

## References

1. A. Sorokin, A. Ostylovsky, A. Borisov, V. Ivanov, D. Makarov, D. Kharlamov. Assessment volume of a forest biomass by the attenuation of the navigation signals the frequencies L1. E3S Web of Conferences 149, 0 0 (2020) https://doi.org/10.1051/e3sconf /20201490 RPERS 2019 20 6 2006
2. V. Podoprigora,, A. Sorokin, D. Makarov, D. Kharlamov. Crown's anisotropy of a tree and scattering the signals of navigation satellites frequency band L1. E3S Web of Conferences 149, 0 0 (2020) https://doi.org/10.1051/e3sconf /20201490 RPERS 2019 20 5 2005
3. Pat. 2682718 Russian Federation, IPC G01S 19/03 (2010.01) Method for determining the linear attenuation coefficients of navigation spacecraft signals in a forest with coordinate reference / Makarov D.S., Savin I.V., Sorokin A.V., Fomin S.V. ., Kharlamov D.V. No. 2018119448; declared 05/25/2018; publ. 03/21/2019, bull. No. 9.
4. Methodology of forest ecosystem research: methodological guide /Comp. E.N.Pilipenko, Vologda-Molochnoe , RC VGMHA, (2013)
5. O'Hara K.L. Stand structure and growing space efficiency following thinning in an even-aged Douglas-fir stand // Canadian Journal of Forest Research. Vol. 18, Pp. 859–866 (1988).

6. Stakanov V.D. The distribution of organic matter in various parts of Scots pine // Forestry,. No. 4. P.25-33. (1990)
7. Linkevicius E, Kuliesis A, Rohle H, Schroder J, Aleinikovas The Impact of Competition for Growing Space on Diameter, Basal Area and Height Growth in Pine Trees // Baltic Forestry. Vol. 20, № 2 (39). P. 301-313. (2014)
8. Nagimov V.Z., Sal'nikova I.S. Crown structure of pine trees Ural forests and forest management, Ekaterinburg, USFEU, vol. 24, pp. 161–169. (2004)
9. G.P. Karev, Structural models of forest ecosystems. Siberian Ecological Journal. 4 pp. 381-396. (1999)
10. S.N. Novik, A.S. Zavyalov, E.D. Telpukhovsky, Electrophysical and dissipative characteristics of the properties of the elements of a coniferous stand // Izvestiya Vuzov. Physics. **49**, 7, P.57-67. (2006)
11. A.V. Sorokin, V.G. Podoprigora, S.V. Fomin, I.V. Savin, Evolution of the signal characteristics of navigation satellites in the forest. Izvestiya vuzov. Physics. 60. 12/2. pp. 19-22, (2017)

E3S Web of Conferences **223**, 03019 (2020) https://doi.org/10.1051/e3sconf/202022303019
*RPERS 2020*

# Impact of climate change on agricultural crops yield on the territory of the Republic of Armenia

*Varduhi* Margaryan [1,]*, *Gennady* Tsibulskii[2] and *Elena* Fedotova[3]
[1]Yerevan State University, Department of Physical Geography and Hydrometeorology, Faculty of Geography and Geology, Yerevan, Armenia
[2] Siberian Federal University, Institute of Space and Information Technologies, Chair of Systems of Artificial Intelligence, 660074 Kirenskogo St. 26, Krasnoyarsk, Russian Federation
[3]Sukachev Institute of forest SB RAS, FRC KSC SB RAS, 660036 Academgorodok 50, Krasnoyarsk, Russian Federation

**Abstract.** The article assesses the change in air temperature and precipitation, and also examines the impact of climate change on crop yields in the Republic of Armenia. As a source of information, the actual data of agrometeorological observations of the GEO "Center for Hydrometeorology and Monitoring" of the Ministry of Environment of the Republic of Armenia and the National Statistical Service are used. As a result of the study, it turned out that there are trends towards an increase in the average annual values of air temperature and heat provision of crops (in total temperatures above 10.0 ° C). In the studied area, there are no regular changes in the amount of atmospheric precipitation over the year. The impact of climate change will only worsen and lead to various problems in water industry, agriculture, energy, health and other sectors. In the republic in 2000-2018 are observed of both the gross harvest and agricultural crops yield increase, with the exception of the tobacco crop. However, this does not mean that agricultural crops are not affected by climate change.

## 1 Introduction

The climate has a decisive influence on agricultural production. Human activities affect the climate, and this factor is unlikely to weaken in the future. Agriculture also has this impact, which at the same time has significant potential for climate change mitigation. Climate changes in farming can generate transformation of crop growth (development stages and their duration), changes of agro-climatic zone's borders, even it is possible to push out the some crops from this area or import of new types of crops [7].

Agriculture is the sector of the economy that is most dependent on the climate. It should be noted that agriculture is one of the main sectors of the Armenian economy, it accounts for about 20% of GDP, and together with the processing of agricultural products - 26%. Moreover, about 60% falls on crop production. On the other hand, risks to food security are increasing due to the expected adverse effects of climate change. Therefore, the strategic task of agriculture is to improve the country's food security, ensure 75-80% of the production of basic food products, increase competitiveness, ensure sustainable development, and develop a system of preventive and adaptation measures to climate change.

Given the above, the work was aimed at studying and analyzing climate change in the study area, discussing and assessing the vulnerability of agricultural crops to climate change, reducing the negative effects of climate change and crops adaptation.

## 2 Study areas, Data and Methods

The information and theoretical basis of this study are current researches in the field of climate change, long-term development programs, government projects and decisions, reports [1, 4–6]. As the source of information actual data of agrometeorological observations of «Hydrometeorology and Monitoring Center» SNCO of the Ministry of Environment of the Republic of Armenia and the National Statistical Service are used.
The Republic of Armenia is located in the northeast of the Armenian Highlands, at the border of the Caucasus and Western Asia. In the north, Armenia borders with Georgia, in the east – Azerbaijan, in the west and southwest – Turkey, and in the south – Iran. The territory of the Republic of Armenia covers 29743 $km^2$. Armenia is a mountainous country: 76,5% of the territory is in the altitudes of 1000-2500 m above sea level. According to the 2006 Land Balance of Armenia, agricultural land accounts for 69 %, forest – 11,5 %, specially protected areas of nature – 12,4 %, wetlands – 0,9 %, settlements, industries, communications, transport and utility infrastructure – 5,4 %, and other land types – 1,7 % of the territory of Armenia [2].
Armenia is a country of climatic contrasts: due to intricate terrain, even in a small territory, a wide variety of climates can be found. The country has almost all types of climate, from arid subtropical to cold high mountainous climates. The average annual ambient air temperature is 5,5°C. The average annual temperature is below zero in altitudes above 2500 m. The summer is temperate: the temperature at the end of July is 16,7°C. The recorded absolute highest temperature is 43,7°C. Winters are cold. January is the coldest winter month, with an average temperature of -6,7°C. The recorded absolute lowest temperature is -42,0°C. Winters in the northeastern and southeastern parts of the country are temperate. The average annual precipitation amounts to 592 mm.
As a methodological basis used are: statistical analysis, correlation methods, modelling.

## 3 Results and discussion

Climate has a direct impact on an agricultural crops growth and yield. Among climatic conditions, temperature is an especially important factor in the viability of crops. The total value of active temperatures in the growing season in general characterizes the probability of growing crops in a given territory, as well as, in particular, the average period and duration of crops phases during the vegetation period. For example [9], grape vegetation begins in spring with sap flow, when the average daily air temperature exceeds 8–10°C and ends in autumn with an average daily temperature below 8–10°C. In 15–25 days after sap flow, when the average daily temperature passes through 11–12°C, the buds open, flowering occurs 35–57 days after the buds open, when the average daily temperature is 18–21°C.
Studies show that according to the actual monitoring data, mainly in the area observed upward trend in average annual air temperature values (fig. 1, a, b).

https://doi.org/10.1051/e3sconf/202022303019

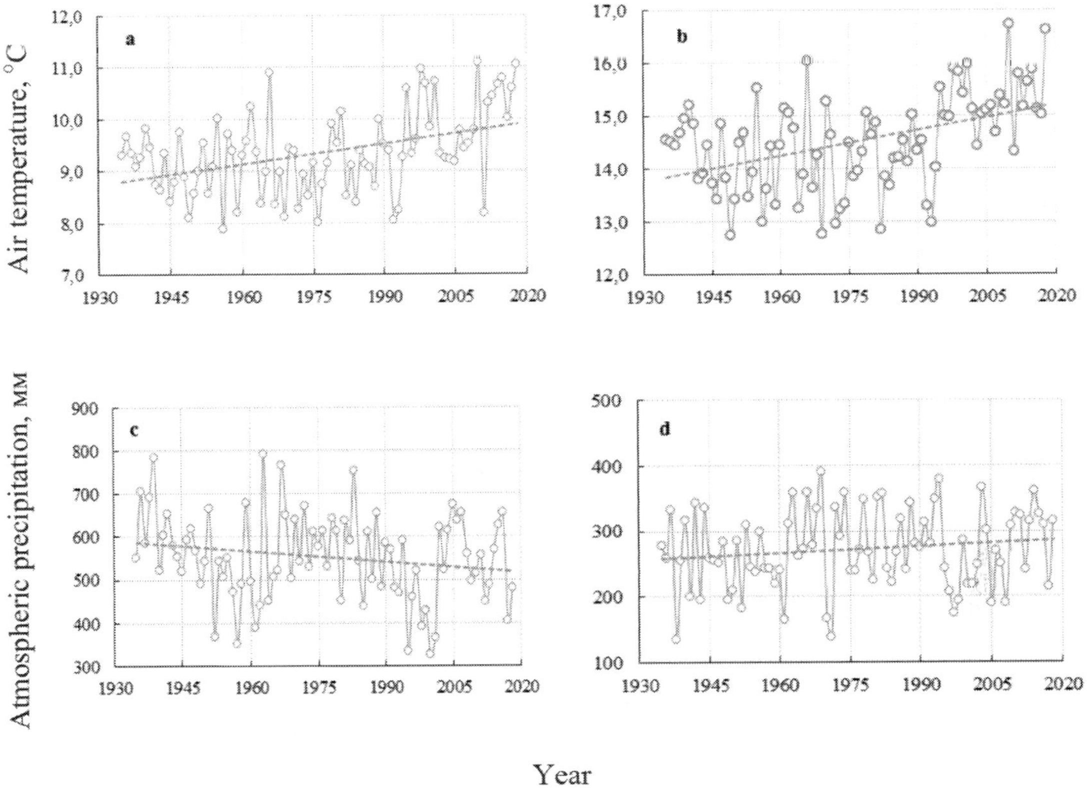


**Fig. 1.** Long-term changes of average air temperature (a, b) and atmospheric precipitation (c, d) for year for stations Odzun (north of Armenia, 1105 m above s.l.) (a, c) and Meghri (south of Armenia, 627 m above s.l.) (b, d).

It should be noted, that the maximum rate of temperature increas is observed in the summer months. In the republic also a tendency of agricultural crops the heat provision increase (the sum of temperatures above 10,0 °C) is observed [9]. Heat provision of agricultural crops is the amount of heat needed to complete the annual cycle of a crop development. It is expressed with the sum of average daily temperatures (the amount of biological temperatures), which accumulate from the beginning of vegetation to a crop maturity. Heat provision has certain effect on a sugar accumulation in fruit, and thus the variety choice, and specialization in viticulture.

With regard to precipitation, the area under study does not show regular changes in the amount of precipitation during the year. For annual precipitation, for different areas both negative and positive trends are observed (Fig. 1, c, d). However, it should be noted that a significant decrease in precipitation in summer is mainly expected.

According to [3], over the period of 1929-1996, the annual average temperature increased by 0,4°C, during 1929-2007 – by 0,85°C, during 1929-2012 – by 1,03°C, and during 1929-2016 – by 1,23°C. The ambient air temperature change has had different trends in different seasons. During the period of 1966-2016, the average summer temperature increased by about 1,3°C, moreover, over the last century, extremely hot summers were observed in Armenia within the last 20 years. During the period of 1935-1996, the average annual precipitation decreased by 6% and in 1935-2016 – by about 9%. The spatial distribution of

precipitation is quite irregular. During the period of 1935-2016 the climate in the north, south and central regions of the country has become more arid, while precipitations have increased in the Shirak plain, in the Lake Sevan basin and in the Aparan-Hrazdan regions.
The results of the CCSM4 Global Climate Model used and reported in the NC3 were reviewed for the projection of changes in air temperature and atmospheric precipitation, as well as high resolution METRAS (12x12 km) regional climate model was applied [3]. The average annual temperature across the territory of Armenia is projected to increase by up to 1,6°C by 2040, by 3,3°C – by 2070 and by 4,7 °C – by 2100, relative to the baseline annual average (5,5°C) for 1961-1990. As for atmospheric precipitations, these are projected to decline by up to 2,7% by 2040, 5,4% – by 2070 and 8,3% – by 2100, relative to the baseline annual average (592 mm) for 1961-1990. It should be noted that there are significant uncertainties in global precipitation assessments due to its high variability and the large number of affecting factors. The issue related to uncertainties in projecting precipitation changes has also been highlighted in the 5th IPCC Report [5].
Therefore, it appears, that the summer season in most regions of the country (hence the vegetation period) is characterized by hot and dry weather conditions. The impact of climate change will only get worse and lead to the generation of various problems in water industry, agriculture, energy, health and other sectors.
On the other hand, it should be noted that the analysis of statistical data shows (table 1), that in the area of the republic in 2000-2018 observed of both the gross harvest and agricultural crops yield increase (fig. 2), with the exception of the tobacco crop. In the Republic, growth of gross harvest and yield of grapes is observed during 2000-2015 [8]. However, this does not mean that agricultural crops are not vulnerable to the impacts of climate change. However, a comprehensive and systematic study of complex events is required.

**Table 1**. Gross harvest and yield capacity of main agricultural crops for the period 2000–2018

| Agricultural crops | Gross harvest, 1 000. t | Yield capacity, centners / ha | Agricultural crops | Gross harvest, 1 000 t | Yield capacity, centners / ha |
|---|---|---|---|---|---|
| Winter and spring wheat* | 245,9 | 24,5 | Legumes* | 4,76 | 23,5 |
| Winter and spring barley* | 132,3 | 20,6 | Tobacco | 1,5 | 26,5 |
| Maize for grain* | 15,4 | 51,9 | Potatoes | 540,1 | 177,6 |
| Oats* | 3,14 | 19,8 | Vegetables | 739,0 | 290,5 |
| Spelt* | 9,62 | 16,9 | Grape | 190,5 | 128,3 |

*weight after processing

In order to assess the impact of climate on the productivity of agricultural crops, numerous correlations between air temperature and various crops were obtained, using the corresponding data of the last 10–15 years. As an example, the paper presents correlations between the average air temperature in July and average crop capacity grain and leguminous crops (fig. 3, a), the average air temperature in March and average crop capacity of vegetable crops (fig. 3, b) for the Armavir marz. Between air temperature and agricultural crops, both directly proportional (fig. 3, a) and inversely proportional (fig. 3, b) correlations were obtained. Moreover, the obtained connections are not close and cannot be used in calculating the yield of agricultural crops. This is because the yield of agricultural crops is formed under the influence of various, including joint weather and climatic factors.

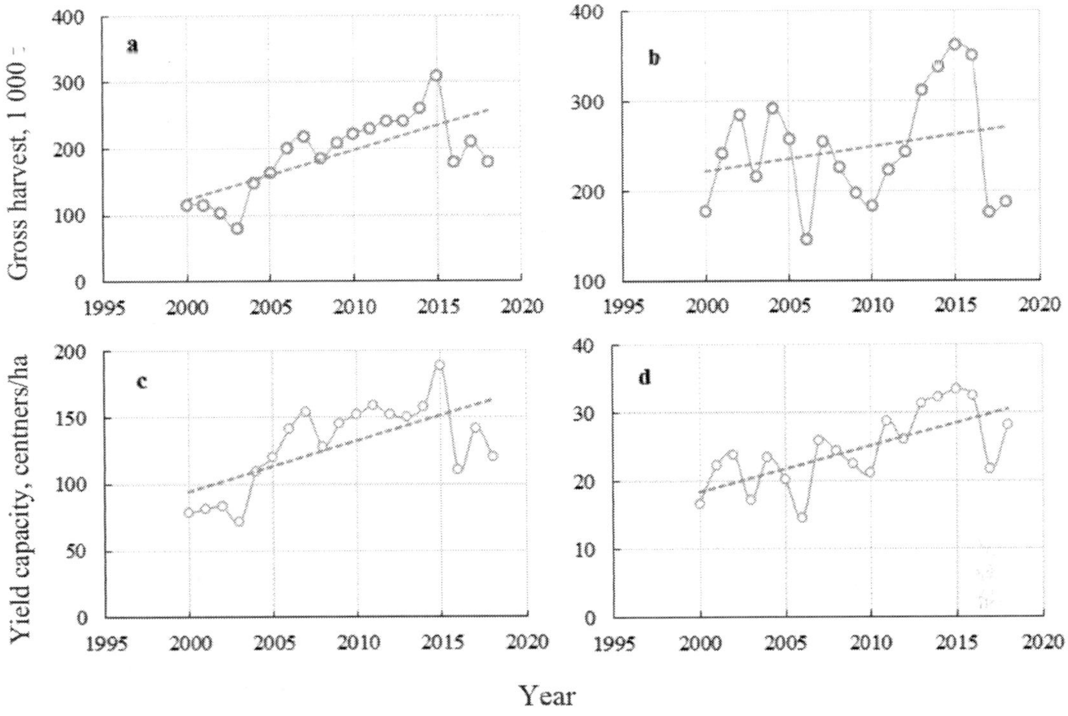


**Fig. 2.** Long-term changes of gross harvest (1 000 t) and yield capacity (centners/ha) for grape (a, c), winter and spring wheat (b, d).

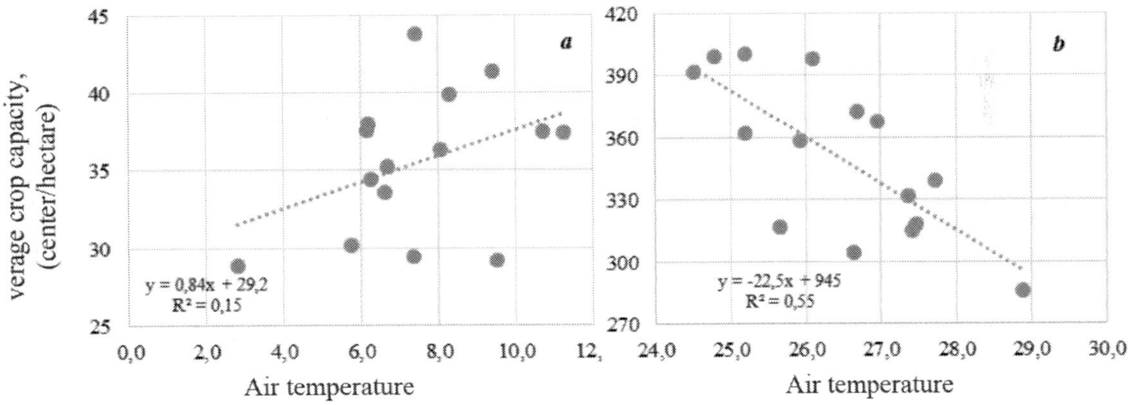


**Fig. 3.** Correlations between the average air temperature in July and average crop capacity of grain and leguminous crops (a), the average air temperature in March and average crop capacity of vegetable crops (b) for the Armavir marz.

## 4 Conclusions

Thus, we arrive to the following conclusions from the studies and recommendations based on them on the territory of the Republic Armenia:
– tendency to increase annual average temperature and of agricultural crops heat provision is observed;
– there are no regular changes in the amount of atmospheric precipitation per year, that is, for the annual amount of precipitation there are both positive and negative trends.;
– in 2010-2018 in the region mainly there was an increase in gross crop harvesting and crop growth.
However, this does not mean that agricultural crops are not vulnerable to the impacts of climate change. There is required implementation of a comprehensive and systematic study of complex events.
– The climate has a direct impact on an agricultural crops growth and yield. The impact of climate change will only get worse and lead to the generation of various problems agriculture and other sectors;
– The agricultural crops are most vulnerable to climate change in the low (irrigated and rain-fed conditions) and middle (dry conditions) zones.
It is necessary to carry out work in the following directions:
– Adaptation of new plants to growing conditions throughout the system running;
– Expansion of irrigated land zones, up to 2000-2200 m, additional irrigation water provided;
– Effective water use technologies, especially in irrigation of fields;
– Drought-resistant and highly productive varieties of agricultural crops selection and implementation of the existing varieties import and localization;
– Investment in infrastructure in the fight against atmospheric adverse phenomena;
– The gradual introduction of the insurance system.

## References

1. *Agriculture sustainability and risk assessment under climate change. Global climate change and risk assessment in agriculture in Russia*. (Saint-Petersburg. 2011)
2. *Armenia's Third National Communication on Climate Change*. (Yerevan, Lusabats. 2015)
3. *Armenia's Fourth National Communication on Climate Change*. (Yerevan, UNDP Armenia. 2020)
4. T.A. Goryanina, A.M. Medvedev The agrarian scientific journal. **12**, 6 (2019)
5. *IPCC: Climate Change 2013: The Physical Science Basis*. Cambridge University Press, Cambridge, United Kingdom and New York, NY, USA (2013)
6. V.G. Margaryan, G.V. Guloyan *MACSUR Science Conference*. (Berlin. 2017)
7. V.G. Margaryan, G.V. Guloyan, H.S. Qocharyan *International conference landscape dimensions of sustainable development: science – planning – governance. ICLDS-2017*. (Tbilisi, Georgia. 2017)
8. V.G. Margaryan, M.S. Mkhitaryan *MACSUR Science Conference*. (Berlin. 2017)
9. V.G. Margaryan, M.S. Mkhitaryan Ukrainian hydrometeorological journal. **21**, 7 (2018)

# Air monitoring system of the Krasnoyarsk Science Center SB RAS

*Valery* Zavoruev[1], *Alexey* Kadochnikov[1], *Alexey* Tokarev[2,*]

[1]Institute of Computational Modeling SB RAS, Krasnoyarsk, Russia
[2]Federal Research Center Krasnoyarsk Science Center SB RAS, Krasnoyarsk, Russia

**Abstract.** The paper deals with the problem of monitoring the surface layer of the atmosphere of the Siberian city of Krasnoyarsk with a population of one million people in order to study the regularities of the formation of a space-time field of particulate matter concentrations in the air. It is shown that such research can be performed only with the help of a system of scientific monitoring of atmospheric air. The Krasnoyarsk Scientific Center SB RAS air monitoring system is a set of special equipment and software designed to collect and analyze objective data on the state of atmospheric air in Krasnoyarsk. The main task of creating and operating the monitoring system is to form an information and analytical basis for research and development aimed at solving the problems of improving the environmental situation in Krasnoyarsk. The article presents the results of almost two years of experience in the operation of the monitoring system and the main results.

## 1 Introduction

The state air monitoring system is regulated by a variety of regulatory documents. Due to this circumstance, the introduction of advanced and generally recognized in many countries technologies and means of monitoring the state of the air environment is practically not implemented in Russia. For example, Hydrometcenter of Russia (Hydrometeorological Research Center of Russian Federation) still measures suspended matter, while many years of international experience show that it is necessary to control particulate matter. It has been shown that it is particulate matter that have a very negative effect on human health [1, 2].

At the level of the constituent entity of the Russian Federation, a regional environmental monitoring system has been created, the operator of which is a subordinate institution of the Ministry of Ecology of the Territory. As part of this system in Krasnoyarsk and the suburbs, there are nine automated observation posts (AOP), where particulate matter concentrations are measured. However, the information received cannot be interpreted in an accessible and understandable form for the population due to bureaucratic reasons prescribed in the regulatory legal acts of Hydrometcenter of Russia.

The residents of Krasnoyarsk are particularly concerned about the smog phenomena, which the people and the media call "black sky". The intensity of the smog is determined by

* Corresponding author: oleg@icm.krasn.ru

the concentration of particulate matter. In this regard, it becomes necessary to study the regularities of the formation of the spatio-temporal field of concentrations of particulate matter in the surface layer of the atmosphere of the city of Krasnoyarsk. For such study of the air environment over the territory of one million city, nine AOP equipped with dust meters are clearly not enough. This problem can be solved by installing relatively cheap air monitoring stations that measure PM concentration and some meteorological parameters [3].

The above circumstances have necessitated the development of a system for scientific monitoring of the air environment. Systems of this type are now being created in many regions, countries and cities of the world [4-8].

## 2 Subject of research

### 2.1 Air monitoring system

The air monitoring system of the Krasnoyarsk Scientific Center of the Siberian Branch of Russian Academy of Sciences (hereinafter referred to as the System) is a set of special equipment and software designed to collect and analyze objective data on the state of atmospheric air in Krasnoyarsk, which, firstly, includes certified and calibrated automatic stations for monitoring atmospheric air pollution (monitoring stations), recording the characteristics of the atmosphere and transmitting this information to a centralized database via a cellular network. Secondly, there are special software tools such as embedded software of monitoring stations and a web interface for their administration, a subsystem for storing registered data and software interfaces for their loading through web services, tools for operational data processing and visualization [9].

The main task of creating and operating the System is to form an information and analytical basis for scientific research and development aimed at solving the problems of improving the environmental situation in Krasnoyarsk.

### 2.2 Measured indicators

The System contains two blocks of information: meteorological characteristics and data on air pollution.

*Meteorological characteristics* are temperature (in degrees Celsius), pressure (mm Hg), relative humidity (percent).

*Air pollution data* is the concentration of particulate matter (PM) with a diameter of less than 10 microns (PM10) and less than 2.5 microns (PM2.5) in the ambient air.

### 2.3 Air Quality Index

Instant AQI air quality index is calculated based on measurements of suspended particles concentration in the System.

The Air Quality Index (AQI) is a widely used worldwide tool for providing information on ambient air pollution in a simple and visual form. In the System, it is currently calculated based on the concentrations of particulate matter (PM10 and PM2.5), which are recalculated (normalized) by piecewise linear interpolation to a single scale of pollution levels (from 0 to 500 units). For each class of this scale, a color designation is also introduced. The main idea of the AQI is to present data on air pollution in terms of human health impact levels.

Along with the basic indicator – the air quality index AQI – its modifications are widely used in practice: NowCast AQI, Asian AQI, Instant AQI (InstantCast AQI), etc. They use revised calculation formulas, in which the averaging of suspended solids concentrations is not performed in 24 hours as in the case of AQI, but in a shorter period – 12 hours (NowCast AQI) or 1 hour (Instant AQI). The result is more objective operational information about air quality; that is why such modifications are recommended for use in public warning systems. The System provides the Instant AQI (instantaneous, operational AQI).

The color-coded AQI index scale shows how air pollution affects human health; on its basis, recommendations were formulated for the population to stay in the open air.

## 3 Materials and methods

The system is based on many years of experience in scientific research and development in the field of solving problems of environmental monitoring and assessment of the state of the natural environment by a team of authors from the Institute of Computational Modeling SB RAS (ICM SB RAS) [10].

The software and technological platform for development is the Geoportal of the ICM SB RAS, including its subsystem "Operational monitoring data".

23 certified CityAir air monitoring stations developed by a group of companies from the Novosibirsk Technopark and the Skolkovo Innovation Center are used as the basic equipment of the air monitoring system of the Krasnoyarsk Scientific Center of the SB RAS.

To assess the data coming from the CityAir air monitoring stations, information from four AOP is used on atmospheric air pollution from the Regional Departmental Information and Analytical System of Data on the State of the Environment of the Krasnoyarsk Territory, which is operated by the Regional State Budgetary Institution Center for the Implementation of Environmental Management and Protection Measures the environment of the Krasnoyarsk Territory". This work is carried out within the framework of intensive information and technical interaction and cooperation, with the support of the regional Ministry of Ecology.

The developed software for the environmental monitoring system structurally consists of a set of interrelated components. The core of the development is a service-oriented data transmission system and a geospatial database, which receives all the collected information. A set of web interfaces, including those adapted for mobile devices, has been created to present data to end users. Data visualization is carried out in several forms: on a map and using a specialized table (Fig. 1).

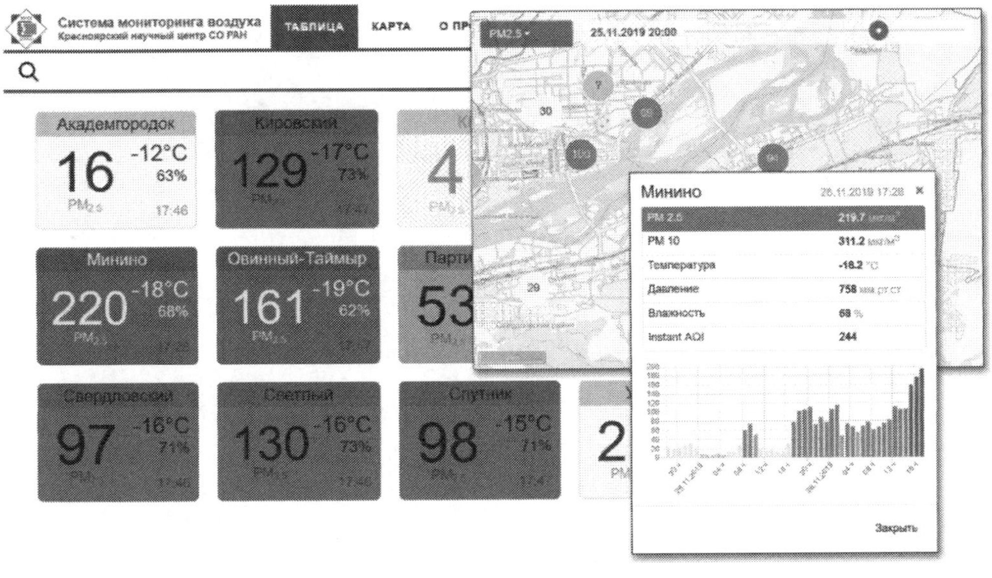

**Fig. 1.** Air monitoring system user interface.

A tabular view of monitoring posts is shown in Fig. 1 on the left side. All posts are displayed as separate "cards" with the ability to search by title and sort. The card contains up-to-date data on air pollution and meteorological information. The color scheme reflects the degree of hazard according to the AQI scale. By selecting a separate post, the user opens an extended view of data with the dynamics of changes in indicators for the last 2 days or a month.

The cartographic presentation of the data (in Fig. 2 on the right) provides display of the hourly averaged pollution values at each observation point, with the ability to scroll through the values for the last 3 days.

## 4 Results and discussion

The system has been operating steadily since October 2018. The advantages of the System include the fact that it stably operates at temperatures of minus 35-40 °C. Under these conditions, measurements on the edge ALP are not carried out for technical reasons. Thus, using the System, for the first time, the concentration of PM2.5 in the atmosphere of the city of Krasnoyarsk was determined during the period of severe frosts [5].

In addition, on the basis of the data obtained using the System, studies are carried out on the regularities of pollution during anticyclones and the effect of forest fire smoke, PM2.5 concentration fields are studied depending on the relief of the city territory and the proximity of the non-freezing Yenisei River.

A comparative analysis of the data for 4 winter months was carried out from 01.11.2019 to 29.02.2020. Graphs of daily average values averaged over all 7 posts of the regional system of the Ministry of Ecology of the Territory (dashed red graph) and averaged over all 16 monitoring stations of the Federal Research Center of the KSC SB RAS (solid blue graph) are shown in Fig. 2.

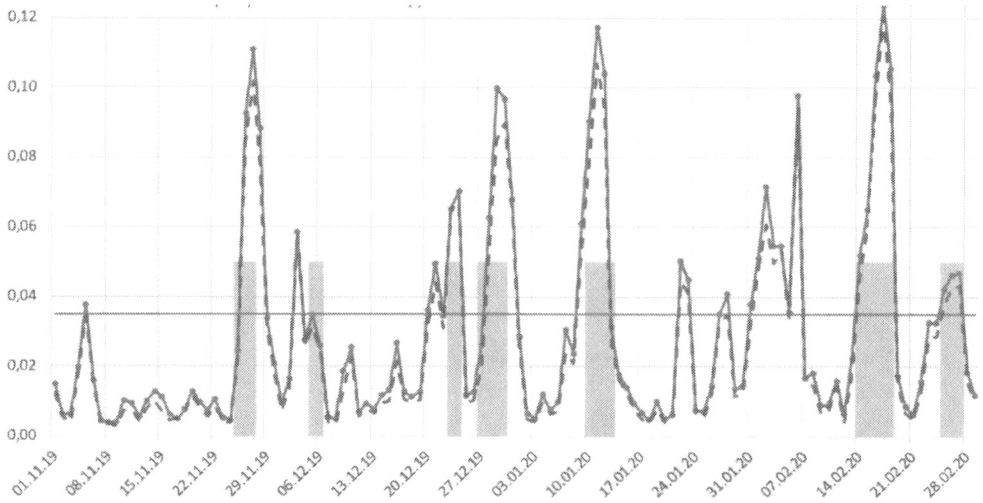


**Fig. 2.** PM2.5 particulate matter concentrations (mg/m³), averaged over monitoring stations in Krasnoyarsk from November 2019 to February 2020. Here are the periods of unfavourable weather conditions announced by state services (gray blocks), the level of average daily maximum allowable concentrations (green line), monitoring data from two independent city monitoring systems (dashed red & solid blue graphs).

 https://doi.org/10.1051/e3sconf/202022303020

This presented graph also shows the days on which the mode of unfavourable meteorological conditions (UMC) was announced, this was 7 times during the period under consideration, and the standard for the average daily maximum permissible concentration PM2.5 = 0.035 mg/m$^3$.

The mean values of PM2.5 concentrations at the monitoring stations for this period are shown in Fig. 3. Analysis of the presented spatial distribution of PM2.5, taking into account the inhomogeneity of the relief, suggests that the pollution of the city's atmosphere in the winter period is primarily due to autonomous heat supply sources (stove heating).

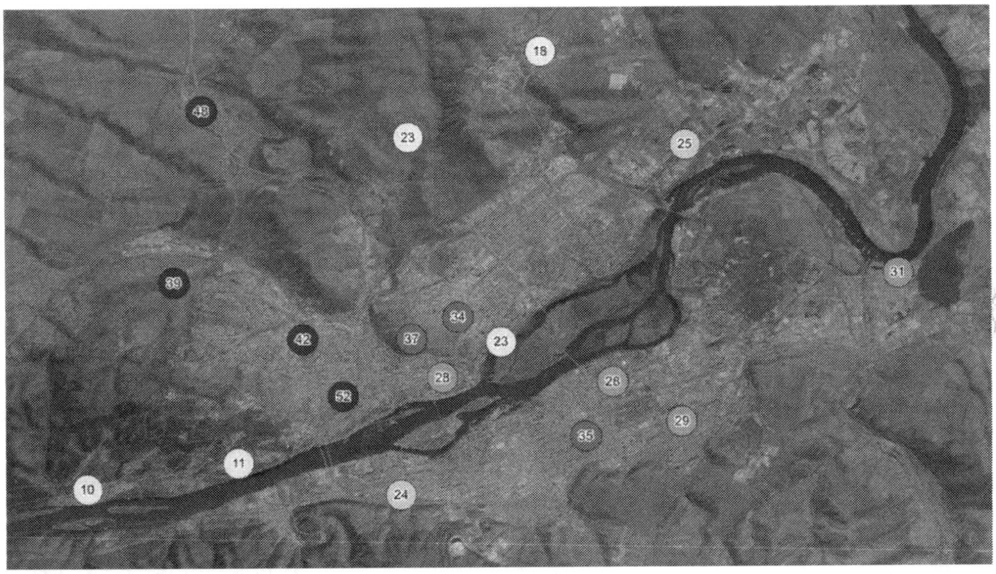


**Fig. 3.** Average concentrations of particulate matter PM2.5 (μg/m$^3$) in winter (for 4 months, from November 2019 to February 2020) at monitoring stations in Krasnoyarsk.

## 5 Conclusions

Within the framework of the presented interdisciplinary research, a development was designed and implemented based on solving the problems of studying the characteristics of the natural environment of a million-strong city and creating information and computing technologies for the development of an effective service-oriented analytical system.

As a result, a network of scientific research monitoring of atmospheric pollution in a large industrial city was formed, based on a large number of spatially distributed operational monitoring stations. This network provides capabilities for detailed spatial and temporal data analysis. For example, it is possible to determine the centers of the initial appearance of pollutants, to trace the dynamics of the spread of pollution across the territory at the level of individual microdistricts of the city. Such data can help to identify sources of pollution, to reveal the characteristic tendencies of its distribution.

The preliminary results of processing the obtained information show a significant effect of furnace heating on the pollution of Krasnoyarsk. The unevenness of the relief, along with the speed and direction of the wind, largely determines the characteristics of the spread of pollution in the city.

Further research in this area is needed to obtain reliable results, analyze the pollution situation and justify proposals for regional authorities.

## References

1. Song C., He J., Wu L., Jin T., Chen X., Li R., Ren P., Zhang L., Mao H. Environ. Pollut. **223**, 575 (2017)
2. S. Gulia, S.M. Shiva Nagendra, M. Khare, I. Khanna. Atmospheric Pollution Research, **6** (3), 286-304 (2015)
3. Changqing Lin et al. Atmospheric Environment. **227**, 117410 (2020)
4. M. Ghodousi, F. Atabi, J. Nouri, A. Gharagozlou. Polish Journal of Environmental Studies, **26** (2), 593-603 (2017)
5. Xuezhen Qiu et al. Frontiers of Environmental Science & Engineering, **9** (6), 1056 (2015)
6. T. Savu, B.A. Jugravu, D. Dunea. REVISTA DE CHIMIE, **68** (4), 796 (2017)
7. L. Hu, P. Yue, M. Zhang, J. Gong, L. Jiang, X. Zhang. Earth Science Informatics, **8** (3), 511 (2015)
8. Hua Wang et al. Journal of Environmental Sciences, **29**, 178 (2015)
9. Kadochnikov A.A., Tokarev A.V., Zavoruev V.V., Yakubailik O.E. IOP Conference Series: Materials Science and Engineering. **537** (6), 062052 (2019)
10. Yakubailik O.E., Kadochnikov A.A., Tokarev A.V. Optoelectronics, Instrumentation and Data Processing, **54** (3), 243 (2018)

# Analysis of temperature inversions during periods of adverse weather conditions in Krasnoyarsk in the winter period of 2019-2020

*Alexander* Dergunov [1*], and *Oleg* Yakubailik [2]

[1]Federal Research Center Krasnoyarsk Science Center of the SB RAS, Krasnoyarsk, Russia
[2]Institute of Computational Modelling SB RAS, Krasnoyarsk, Russia

**Abstract.** The work is devoted to the search for relationships between the pollution of the atmosphere of Krasnoyarsk by particulate matter and temperature inversion – an increase in temperature with height in the surface layer of the atmosphere. The research is based on reanalysis data of the NASA GFS meteorological model for air temperature at different altitudes of the atmosphere and the results of measurements of concentrations of particulate matter in the air monitoring system of the FRC KSC SB RAS, as well as information about officially declared periods of adverse weather conditions. The results obtained allow us to conclude that there is a high degree of correlation between these values, and that it is possible to use the GFS model data to predict the environmental situation.

## 1 Introduction

The problem of air pollution in large cities in the world is acute. Numerous studies show that there is a link between increased concentrations of airborne particulate matter (PM) and deterioration in human health in metropolitan areas [1]. A sharp increase in the concentration of pollutants occurs during the period of adverse weather conditions, which contribute to the accumulation of industrial and automobile emissions. [2]. However, not only the anthropogenic factor, but also natural processes contribute to the creation of unfavorable environmental conditions. For example, temperature stratification of the lower atmosphere can cause temperature inversion, that is, a layer of warm air that limits the vertical dispersion of harmful impurities in the atmosphere [3].

Temperature inversions arise because layers of greater or lesser thickness are located in the surface layer of the atmosphere at different heights, in which the decrease in temperature greatly slows down, stops, or vice versa, instead of decreasing temperature with height, it increases. Their properties are the height of the lower and upper boundaries, vertical thickness (layer thickness) and intensity (inversion value). Temperature inversions are divided into 3 types: surface (the lower boundary is located at ground level), elevated (the lower boundary is located at a certain height from the surface) and free atmosphere inversions (the height can vary greatly) [4, 5, 6].

* Corresponding author: alexdergunov@icm.krasn.ru

According to the Ministry of Ecology and Rational Nature Management of the Krasnoyarsk Territory, in Krasnoyarsk, from January 2019 to February 2020, 8 periods of adverse weather conditions were established: in 2019 - from 8 to 13 February, from 26 to 28 November, from 5 to December 6, from December 23 to December 24 and from December 27 to 30; in 2020 - from 10 to 13 January, from 14 to 18 February and from 25 to 27 February [7]. These periods are characterized by low wind speeds and airborne particulate matter (PM) values well above the daily average maximum permissible concentration (0.035 mg/m$^3$).

Figure 1 shows an image of the study area, namely the Krasnoyarsk city and its surroundings. It is a city with a population of over 1 million people, located along the Yenisei River at a distance of about 40 kilometers.

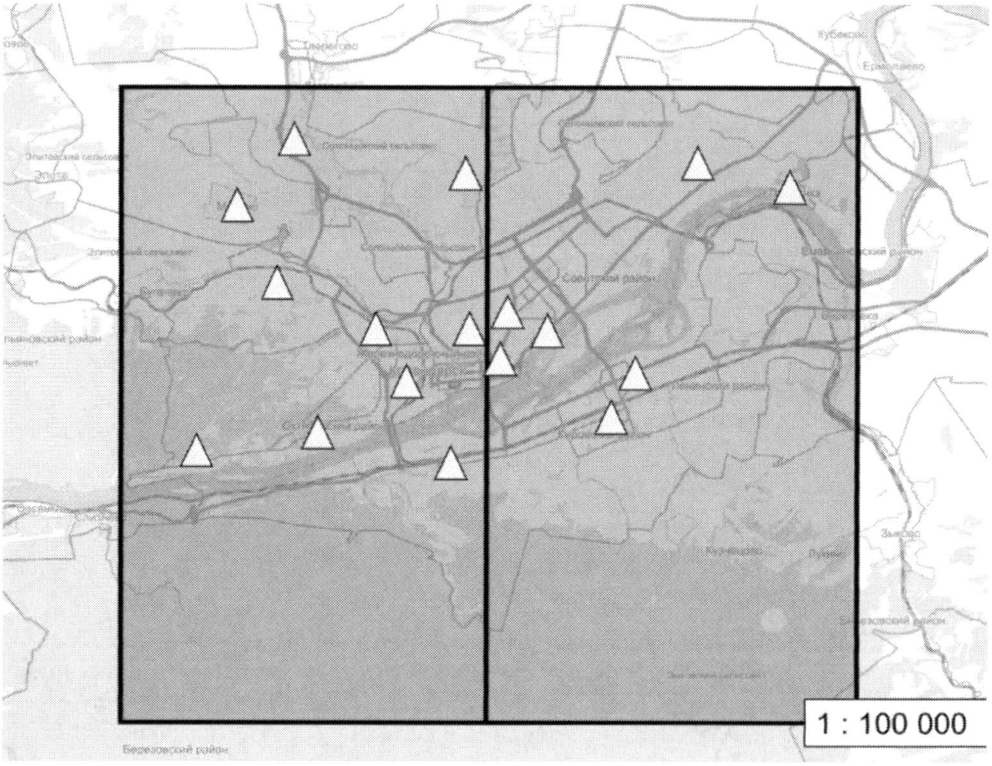


**Fig. 1.** Study area: Krasnoyarsk city and its surroundings.

In Figure 1, yellow triangles denote 17 ground-based ecological monitoring stations of the Federal Research Center of the Krasnoyarsk Science Center Siberian Branch of Russian Academy of Sciences (FRC KSC SB RAS), two cells corresponding to the regular grid of the NASA GFS (Global Forecast System) weather forecast model are shown in light red.

In this work, the meteorological conditions in the city of Krasnoyarsk in the winter period of 2019-2020 were studied using data from ground-based monitoring posts and meteorological information from the GFS weather forecast model.

## 2 Materials and methods

The GFS model is a numerical weather prediction system containing a global computer model and variation analysis performed by the United States National Weather Service (NWS). This model is a collection of four separate models: atmosphere, ocean, land / soil, and sea ice. This is one of the world's most famous meteorological models. Global data analysis and forecasting is carried out 4 times a day. The weather forecast is available up to 16 days ahead [8]. Since January 2015, the horizontal resolution of the GFS model data is 0.25 degrees (about 25 km).

GFS model data is in *.grib2 format. This data format is standardized by the World Meteorological Organization (WMO) and is designed to store historical and forecast weather data [9]. Each individual file contains more than 500 layers of different meteorological information at more than a hundred vertical levels.

In this work, we used actual data from the analysis of air temperature at the ground level and at heights of 750 and 1500 m [10].

The processing of the initial data of the GFS model consisted of several stages. At the first stage, the resulting files were cropped according to the specified coordinates corresponding to the city of Krasnoyarsk, and layers containing information about the air temperature at a certain height. Then, the data trimmed by coordinates and layers was converted into tabular files in *.csv format. These procedures were performed using special scripts written in Python using the wgrib2 program. The wgrib2 program is specially designed by the GFS model data producer for reading and converting *.grib2 data [11].

At the second stage of processing, the contents of the resulting *.csv files were combined into one large common *.csv file using a small program written in the C language. Further processing and analysis of data was carried out in Microsoft Excel software.

The inversion layers were identified by obtaining the difference between the temperature data at different heights. Temperature inversion was defined as the case when the difference gave a negative value. For example, if a negative value was obtained between the temperature difference on the surface and at an altitude of 750 m, the inversion was considered surface or elevated, if between 750 and 1500 m - an elevated or high-altitude inversion (inversion of the free atmosphere). If the surface/elevated and high-altitude inversions were recorded simultaneously, then this was considered a powerful inversion.

Also, the work used data on the concentration of suspended solid particles in the air PM2.5 and wind speed, which were obtained on the geoportal of the ICM SB RAS [12]. The daily average PM2.5 data were averaged over all available observation sites in the city to obtain the daily average PM2.5 concentration values for the entire city of Krasnoyarsk.

## 3 Results and discussion

An archive of data was obtained for the period from January 1, 2019 to April 30, 2020, containing information on air temperature at surface level and heights of 750 and 1500 meters, average daily PM2.5 particle concentrations and information on wind speed for the city of Krasnoyarsk.

Figure 2 shows the variations in PM2.5 concentration and wind speed over the city of Krasnoyarsk from November 1, 2019 to February 29, 2020.

 https://doi.org/10.1051/e3sconf/202022303021

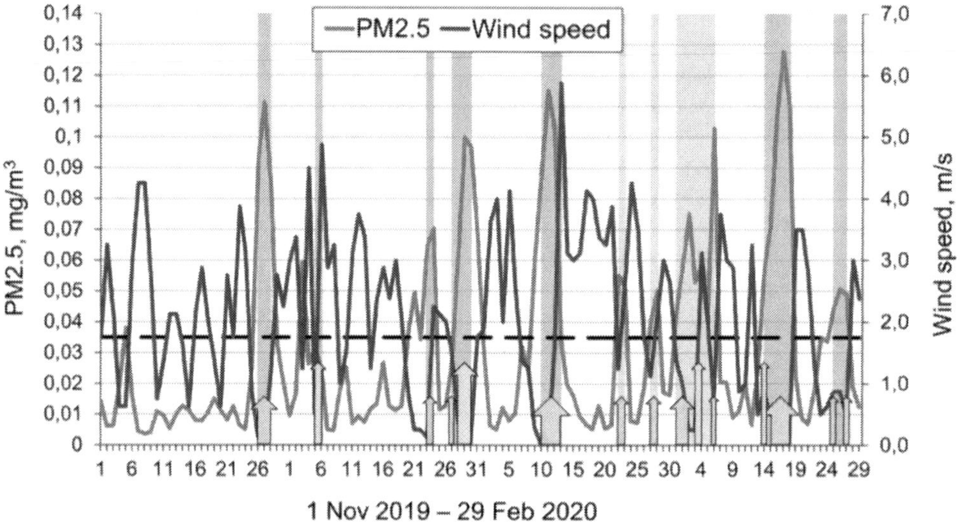


**Fig. 2.** Changes in the average daily values of PM2.5 concentration and wind speed in the city of Krasnoyarsk in winter 2019-2020. The dotted line indicates the value of the average daily maximum permissible concentration of PM2.5, equal to 0.035 mg/m$^3$.

Figure 2 shows the following data:

- the blue vertical stripes indicate the official periods of adverse weather conditions in the city of Krasnoyarsk, and
- the orange vertical stripes indicate the periods identified on the basis of data analysis when the daily average PM2.5 concentration values significantly exceeded the maximum permissible concentration, which is 0.035 mg/m3 in accordance with Russian legislation.

Arrows are plotted on the abscissa:

- green arrows indicate periods when surface or elevated temperature inversion was present;
- blue arrows indicate periods when powerful inversion was present.

Here in the figure 2, three features stand out. First, the relationship between the values of the average daily PM2.5 concentration and wind speed. Whenever PM2.5 concentration values significantly exceeded the maximum permissible value, wind speed was extremely low or absent. Second, in all periods when PM2.5 concentration values significantly exceeded the daily average maximum allowable value, temperature inversions were observed over the city. Third, in addition to the official regimes of adverse weather conditions, characterized by increased values of PM2.5 concentration, in the period from November 1, 2019 to February 29, 2020, 3 more periods were identified with a similar nature of meteorological conditions and the level of air pollution. This means that the methods used by official government services to predict unfavorable weather conditions are imperfect and sometimes do not determine the upcoming periods of high air pollution.

The table shows the results of the analysis of the atmosphere of the city of Krasnoyarsk in the winter period of 2019-2020 according to the obtained data.

Table 1. Data on the state of the atmosphere of the city of Krasnoyarsk and its surroundings in the winter period of 2019-2020

| Date | PM2.5, mg/m³ | Wind speed, m/s | Temperature inversion |
|---|---|---|---|
| 1-5 Jan 2019 | 0,029-0,175 | 0,125-2,75 | All days. Surface or elevated |
| 11-13 Jan 2019 | 0,048-0,073 | 0,875-1,125 | January 12 and 13. Surface or elevated |
| 23-25 Jan 2019 | 0,057-0,106 | 0-0,75; 2,4 – 25 Jan | January 24 and 25. Surface or elevated |
| 08-13 Feb 2019 | 0,064-0,165 | 0,25-0,8<br>11,625 – 13 Feb | All days. Surface or elevated |
| 26-28 Nov 2019 | 0,085-0,111 | 0-0,125 | November 26 and 27. Surface or elevated |
| 05-06 Dec 2019 | 0,025-0,032 | 0,5; 4,9 – 6 Dec | All days. Powerful |
| 23-24 Dec 2019 | 0,065-0,07 | 0,125-2,25 | All days. Surface or elevated |
| 27-30 Dec 2019 | 0,026-0,099 | 0-1,5 | All days. Powerful |
| 10-13 Jan 2020 | 0,035-0,115 | 0-1,5; 5,9 – 13 Jan | All days. Surface or elevated |
| 22-23 Jan 2020 | 0,052-0,055 | 1,25-2,6 | All days. Surface or elevated |
| 27-28 Jan 2020 | 0,04-0,05 | 1,125-2,125 | All days. Surface or elevated |
| 31 Jan - 06 Feb 2020 | 0,039-0,075 | 0,25-3,125 | All days. January 31, February 1, 2, 5, 6 - surface or elevated. February 3 and 4 February - powerful |
| 14-18 Feb 2020 | 0,057-0,127 | 0-0,87 | All days. Surface or elevated |
| 25-27 Feb 2020 | 0,045-0,05 | 0,5-0,87 | All days. Surface or elevated |

In this table, eight out of fourteen periods are the official periods of adverse weather conditions in the city. The remaining six periods are identified in the process of data analysis. All the periods presented are characterized by daily average values of PM2.5 concentration exceeding the maximum permissible concentration, as well as low wind speed in the city and the presence of temperature inversions. In four cases, the wind speed sharply increased on the last day of the period, while the PM2.5 concentration on that day dropped significantly. The prevailing type of detected inversions is surface or elevated. In the official period from 5 to 6 December 2019, the average daily concentration of PM 2.5 in the city did not exceed the daily average value of the maximum permissible concentration.

## 4 Conclusions

As a result of this work, a relationship was shown between high values of the average daily PM2.5 concentration and wind speed in the city of Krasnoyarsk. The presence of temperature inversions during periods of adverse weather conditions was revealed. Data analysis revealed an additional six periods when the daily average PM2.5 concentration significantly exceeded the maximum allowable value, in addition to the official eight periods of adverse weather conditions. The predominant type of identified temperature inversions is surface or elevated.

Analysis of meteorological data of high spatial resolution, for example, the GFS model, helps to solve problems related to the study of the lower atmosphere and its pollution, and can also play an important role in more accurately identifying periods of adverse weather conditions and their forecasting.

## References

1. Y. J. Kaufman, D. Tanré, O. Boucher, Nature **419**, 6903 (2002)
2. A.I. Eremkin, *Regulation of emissions, pollutants into the atmosphere*, Assoc. of Constr. Univers. (2001)
3. K. Smith, *Principles of applied climatology,* Hydromet. (1978)
4. L. T. Matveev, *Physics of the atmosphere,* Hydromet. (2000)
5. E. Y. Bezuglaya, *What an industrial city breathes,* Hydromet. (1991)
6. N. P. Krasnenko, *Acoustic sounding of the atmospheric boundary layer*, «Vodolei» (2001)
7. The Ministry of Ecology and Rational Nature Management of the Krasnoyarsk Territory Retrieved from: http://www.krasecology.ru/Nmu
8. The Global Forecast System (GFS) Documentation from: https://www.emc.ncep.noaa.gov/emc/pages/numericalforecast_systems/gfs.php
9. WMO GRIB Edition 2 Documentation Retrieved from: https://www.wmo.int/pages/prog/www/WMOCodes/Guides/GRIB/GRIB2_062006.pdf
10. NCEP GFS 0.25 Degree Historical Archive Retrieved from: https://rda.ucar.edu/datasets/ds084.1/
11. Wgrib2 Utility to read and write grib2 files Retrieved from: https://www.cpc.ncep.noaa.gov/products/wesley/wgrib2/
12. Geoportal ICM SB RAS Retrieved from: http://gis.krasn.ru/blog/

# Comparison of different models for assessing air quality in Krasnoyarsk using satellite data

*Konstantin* Krasnoshchekov[1,*], *Oleg* Yakubailik[2]

[1]FRC KSC SB RAS, Krasnoyarsk, Russia
[2]Institute of Computational Modeling SB RAS, Krasnoyarsk, Russia

**Abstract.** Methods for estimating the atmospheric pollution of Krasnoyarsk by particulate matter based on satellite data on the aerosol optical depth (AOD) are considered. Satellite data from the MODIS MAIAC algorithm with a spatial resolution of 1 km are used together with data from the ground-based $PM_{2.5}$ environmental monitoring stations of the FRC KSC SB RAS research network. A comparative analysis of the relationship between the calculated values of $PM_{2.5}$ obtained from AOD data and ground-based measurements of $PM_{2.5}$ in the summer of 2019 is presented. Various models of the relationship between these parameters were investigated, and a high level of correlation of these values was obtained. The calculated coefficient of determination was about 0.7.

## 1 Introduction

Numerous epidemiological studies show that there is a relationship between the concentrations of particulate matter (PM - particulate matter) in the atmosphere and public health [1-2]. Attention to this issue is constantly increasing, with PM monitoring networks being established in many cities. Monitoring stations usually measure the concentration of three types of particulate matter ($PM_{10}$, $PM_{2.5}$, $PM_1$ - particles of different sizes). $PM_{2.5}$ is considered the most dangerous for human health, since the particle size up to 2.5 microns allows them to enter the human bloodstream through breathing and lead to various diseases.

PM monitoring is based primarily on ground measurements. Networks of monitoring stations have been established today in many cities, but the number of observation points in them, as a rule, is small, these point measurements do not provide detailed information on the spatial characteristics and the distribution of PM in urban areas of interest. The time span for on-site PM measurements also varies greatly depending on the period of operation of the instrument and its functionality. These reasons determine the development of works on PM assessment by satellite remote sensing methods.

The AOD parameter (aerosol optical depth) is a value that characterizes the attenuation of sunlight as it passes through the atmosphere. Particulate matter in the atmosphere (dust, smoke, pollution) absorbs or scatters light. The AOD value depends on the amount of particulate matter that prevents direct sunlight from reaching the ground. This is a dimensionless number that is related to the amount of PM in the vertical column of the

* Corresponding author: krasko@icm.krasn.ru

atmosphere above the observation site. Numerous linear [3], chemical [4], transport [5] and neural [6] models have been developed to determine how AOD correlates with ground-based PM measurements. However, the size distribution of particles, their composition, air humidity, and wind speed significantly reduce the value of the AOD-PM correlation [6].

## 2 Data and measurement

### 2.1 Study area

The city of Krasnoyarsk is actively developing and currently has an area of about 350 $km^2$. It is located in a river valley, stretching 20-25 km along the river and 3-4 km inland, abutting the mountain ranges that frame the valley. The city lies as if in a bowl, at the bottom of which a river flows. These features of the relief, which form the ecological environment of the city, determine the circulation of air masses and the low scattering power of the atmosphere, which leads to the accumulation of pollutants in the city and suburban areas. From the south and west of the city there are forests and hilly terrain, from the north and east the relief is predominantly flat. Like all major cities in the world, Krasnoyarsk is subject to negative impact on the environmental situation. Its deterioration is facilitated by the fact that the city is the largest transport hub in Eastern Siberia; the presence of a number of large enterprises in the metallurgical, engineering and chemical industries aggravates the situation. It is shown in [7] that the concentration of $PM_{2.5}$ in Krasnoyarsk is 64% higher than the average for Russia.

### 2.2 Satellite observation of $PM_{2.5}$

The study of the parameters of aerosols in the atmosphere using specially developed methods and technologies for processing Earth remote sensing data are devoted to many works. Since the launch of the Terra and Aqua satellites about 20 years ago, algorithms have been proposed for obtaining AOD with a spatial resolution of 10 and 3 km per pixel. Later, the method of interpreting the recorded data was improved, algorithms with a higher spatial resolution (1 km) were proposed.

In the course of our work, we used AOD data with a spatial resolution of 1 km, measured using a MODIS spectrophotometer, MAIAC algorithm, product MCD19A2. The MCD19A2 product contains data from the MODIS spectrophotometer installed on the Terra and Aqua satellites. This product was published on May 30, 2018 and contains AOD data since February 1, 2000. MAIAC provides a uniform 1 km grid resolution in the selected projection regardless of the scan angle. MAIAC data has limitations in obtaining information about AOD. There should be no clouds over the area, it should not be covered with snow. Consequently, data on AOD in winter for Krasnoyarsk are not available.

### 2.3 Ground-based observations of $PM_{2.5}$

The research monitoring network of the Federal Research Center "Krasnoyarsk Science Center of the Siberian Branch of the Russian Academy of Sciences" (FRC KSC SB RAS) was selected as the information source for the ground-based measurements of $PM_{2.5}$. In total, data from 13 monitoring stations were used. The FRC monitoring network was deployed in the fall of 2018.

In our study, we used $PM_{2.5}$ measurements from the FRC posts and AOD measurements for the period from June to August 2019. The choice of this time range is due to several factors. First, for summertime data, it is possible to compare ground-based and satellite measurements, which is the subject of research in this work. Second, a preliminary analysis of the data showed that at the indicated time, there were significant variations in air pollution in the city associated with the short-term but strong impact of forest fires in the vicinity of the city. We studied the relationship between AOD and $PM_{2.5}$ concentrations on the Krasnoyarsk city scale. The location of $PM_{2.5}$ ground-monitoring stations is shown in Fig. 1.

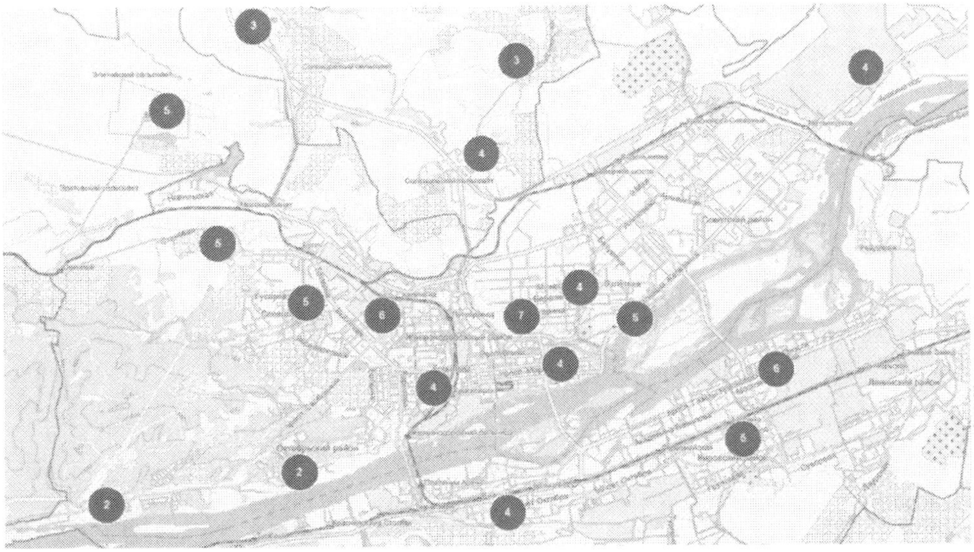

**Fig. 1.** Location of monitoring stations of KSC.

The green circle marks the location of monitoring stations in the city of Krasnoyarsk as of October 2020. Their distribution throughout the city is relatively uniform, but taking into account the features and inhomogeneities of the relief. A sufficient number of ground-based monitoring stations allows one to obtain comparable results with the MODIS MAIAC satellite data used in this work.

It is worth paying special attention to the timing of measurements of ground data. The city monitoring system takes measurements continuously, with a time step of several minutes. At the same time, analysis of the data shows that due to changes in meteorological conditions (for example, wind speed and direction), the situation with city pollution can radically change in a few hours. Therefore, in the framework of this study, we used data averaged over time intervals corresponding to the time of image registration by the Terra and Aqua satellites.

### 2.4 Meteorological observation

Part of the meteorological information required for the calculations was obtained from $PM_{2.5}$ ground monitoring stations (temperature and air humidity). The planetary boundary layer height (PBLH) parameter is derived from the NASA Global Forecast System (GFS) atmospheric model.

### 2.5 $PM_{2.5}$ estimation from satellite data

This paper compares the calculated $PM_{calc}$ data using a linear regression model that includes AOD values, formula (1) and an algorithm that includes AOD data, boundary layer heights, air humidity, aerosol gyroscopic growth coefficient (formula (2)) [8]. The calculated $PM_{calc}$ results were compared with the values measured at ground monitoring stations.

The linear regression equation is

$$PM_{calc}=a \cdot AOD+B \quad (1)$$

where $a$ and $B$ are the linear regression coefficients and are equal to $7.2 \cdot 10^{-3}$ and $5.3 \cdot 10^{-3}$, respectively. The result obtained using the linear model is shown in Fig. 3.

However, the linear model does not take into account such parameters as air humidity RH, boundary layer height PBLH) coefficient of gyroscopic aerosol growth $\gamma$. Therefore, the spatial distribution of PM concentrations was estimated using the refined model:

$$PM_{calc}=[AOD/(A \cdot PBLH)]/\{K \cdot [(1-RH)/(1-RH_0)]^{-\gamma}\} \quad (2)$$

where AOD is the aerosol optical thickness data obtained by satellite methods; RH is the air humidity; PBLH is the height of the atmospheric boundary layer.

The result is shown in Fig. 3. PBLH was obtained from GFS data, RH parameter was measured using ground-based monitoring stations. The $\gamma$ and K values are aerosol characteristics and are derived from [8].

## 3 Results and discussion

Since satellite data for the territory of the city of Krasnoyarsk are not available in winter due to the presence of snow cover, and the network of ground monitoring stations was deployed in the fall of 2018, only the period from June to August 2019 was studied. The data of the MODIS MAIAC algorithm with a resolution of 1 km were used, since the spatial resolution is of great importance on a city scale. A comparison of this data with the less detailed MODIS MOD04 data (10 km) is shown in Fig. 2.

**Fig. 2.** AOD map for the territory of Krasnoyarsk with its surroundings. Data calculated using MODIS algorithms with a spatial resolution of 1 km (a) and 10 km (b).

During the study period, not a single unfavorable weather conditions mode was introduced. However, the monitoring stations registered the excess maximum permissible concentration. The AOD values were taken over the location of ground monitoring posts, then they were averaged and the average value for the city was obtained. From the obtained value, the PM value was calculated using formulas 1 and 2. The calculated $PM_{calc}$ values were compared with the PM values, which were obtained as an average value from all monitoring posts, taken at 12:00 local time.

The result obtained using the linear model is shown in Fig. 3. The calculated $PM_{calc}$ values using the linear model have a coefficient of determination equal to 0.73. It follows that the

PM data calculated using satellite monitoring have a high agreement with the data measured at ground-based monitoring stations.

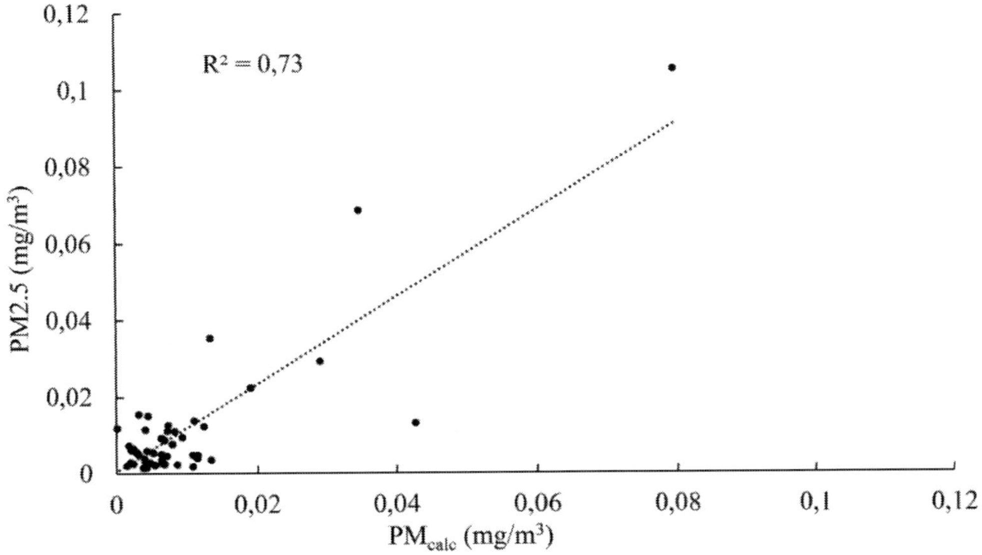


**Fig. 3.** Comparison of PM2.5 received from the posts of KSC and PM and calculated using a linear model ($PM_{calc}$).

Using formula 2, we obtained a coefficient of determination of 0.67, which indicates a fairly good fit with the ground-based $PM_{2.5}$ data (Fig. 4).

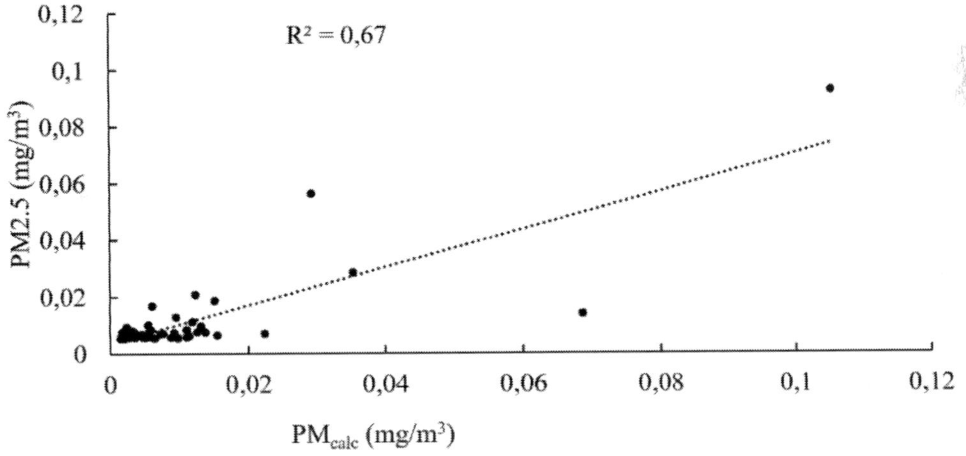


**Fig. 4.** Comparison of PM received from KSC posts and calculated values using remote sensing data

The results obtained using the linear regression model showed a higher correlation with ground-based observations when compared with the values obtained using the model described in [7]. However, in both cases, a sufficiently high correlation with ground-based data was obtained. Using data from environmental monitoring posts together with data of high spatial resolution MODIS MAIAC, it is possible to create a basis for improving the environmental situation of the city.

## References

1. Y.J. Kaufman, D. Tanré, O. Boucher, Nature **419,** 6903, 215 (2002)
2. J. Schwartz, F. Laden, A. Zanobetti Environ. Health Perspect. **110,** 10, 1025 (2002)
3. X.Q. Yap, M. Hashim ACP. 13 3517 (2013)
4. I. Kloog, F. Nordio, B. A. Coull, J. Schwartz, Environ. Sci. Technol **46**, 21, 11913 (2012)
5. P. Gupta, S.A. Christopher J. of Geophysical Research **114**, 14205 (2009)
6. J. Wang, S.A. Christopher Geophys. Res. Lett. **30**, 21, 2095 (2003)
7. C. Lin, L.D. Labzovskii, H.W.L. Mak, J.C. Fung, A.K. Lau, S.T. Kenea, J. Ma Atmos. Environ. 117410 (2020)
8. C. Lin, Y. Li, Z. Yuan, A.K. Lau, C. Li, J.C. Fung Remote Sens. Environ **156,** 117 (2015)

# Cataloging system for Russian satellites remote sensing data: main characteristics and operating experience

*Alexey* Kadochnikov[1], *Alexey* Tokarev[2,*]

[1]Institute of Computational Modeling SB RAS, Krasnoyarsk, Russia
[2]Federal Research Center Krasnoyarsk Science Center SB RAS, Krasnoyarsk, Russia

**Abstract.** A local archive and a system for cataloging satellite data from Russian spacecraft that are received by the FRC KSC SB RAS through the services of the Roscosmos Geoportal have been developed. An automated system for remote sensing data preprocessing has been created, which significantly simplifies access to information, implements search services and web visualization of images registered in the catalog. Practical experience in operating the system has revealed a number of shortcomings of the existing Roscosmos data distribution system. Presented data on the execution of the Roscosmos satellite survey order in 2019 they show that the capabilities of the Russian remote sensing system are very limited, and a significant part of the planned research work has not been implemented due to the lack of ordered satellite data.

## 1 Introduction

The main source of Earth remote sensing (ERS) data from Russian spacecraft is the Roscosmos Geoportal, which is operated by the Scientific Center for Earth Operational Monitoring (NTs OMZ) of Russian Space Systems JSC. It is a web resource that combines the means of searching and viewing, ordering of new and archived remote sensing data from Russian spacecrafts. Data receiving is possible through the web interface of the geoportal or via the ftp protocol.

Several institutes and divisions of the Federal Research Center of the Krasnoyarsk Science Center Siberian Branch of Russian Academy of Sciences (FRC KSC SB RAS) are interested in using data from Russian spacecrafts. Large volumes of remote sensing data that need to be obtained through the Roscosmos Geoportal, as well as an analysis of its capabilities, which revealed a number of fundamental limitations, led to the need to create a local archive of satellite information, access to which is provided to all interested divisions of the organization. A technology was developed for automated processing of data coming from the ftp server of Roscosmos in the form of a specialized service running on the server. The proposed approach made it possible to simultaneously solve a number of discovered technical problems associated with the received ERS data.

* Corresponding author: oleg@icm.krasn.ru

The development of Earth remote sensing systems is currently due to a combination of factors such as an increase in the number of spacecraft and an improvement in their technical characteristics, an increase in the availability of satellite information associated with the development of the Internet: an increase in speed and a decrease in the cost of access, the development of web services and geospatial transmission standards. data [1]. New technologies for automating the processes of receiving and processing Earth remote sensing data, creating super-large archives in specialized data centers have made it possible to organize fundamentally new schemes for working with satellite data, to create a new generation of systems for operational monitoring of the natural environment. Geographic information web technologies, interactive online cartographic systems with direct access to satellite information, provided with the ability to customize various parameters of visualization of geospatial data and build complex analytical queries, have become their key elements.

Technologies of geoinformation web systems and services are dynamically developing at present [2-4]. More than a hundred Russian geoportals, dozens of geoservices, web GIS, cartographic services and data visualizers can be found on the Internet. Including specialized solutions for working with satellite data. Along with the already traditional tasks of simple visualization of terrain maps and mosaics of satellite images, modern systems provide the user with the possibility of flexible adjustment of various parameters of cartographic visualization, services for processing spatial data [5].

## 2 Subject of research

The developed system for cataloging satellite data coming from Russian spacecraft is currently focused on data from the Resurs-P and Canopus-V satellites. The data are centrally loaded for the needs of the departments included in the Federal Research Center of the KSC SB RAS, according to a common request for all departments. For the convenience of the end users of the catalog, the received data is processed and brought to a single format. Now these data are automatically downloaded from the Roscosmos geoportal (https://www.gptl.ru) and are pre-processed in a semi-automatic mode.

Access to images from the Roscosmos geoportal is organized through the ftp server. Data on the geoportal server is stored for a limited time (usually no more than a month), therefore it is necessary to periodically upload them to your storage. Each image contains shapefiles with its borders or borders of each channel separately, xml with the image metadata, jpeg with an overview image and one or more tiff files with raster data with one or more bands. XML metadata format and file name format for each type of spacecraft is different. XML formats also differ for different levels of image processing. The data is prepared and placed on the portal by the operators of the supplier manually, and as practical experience has shown, this leads to a lot of errors, different variants of the data presentation structure.

The disadvantages include the fact that the file names contain the processing time of the operator, although it would be more convenient to store the shooting time in them. Folder names can contain Russian characters, which introduces a number of problems when working with them in Unix systems when converting the encoding.

## 3 Software technologies

To search and navigate the satellite data catalog, ftp access was organized within the network of the Federal Research Center of the KSC SB RAS and a web application for the satellite images catalog was developed using the server programming language PHP,

PostgreSQL DBMS, SQLite3 and the client web interface in the TypeScript programming language using the Angular framework 5 [2, 3].

The developed web application uses a set of automatically generated multiscale "natural color" overview images, as well as overview images with the NDVI index. Specially prepared multiscale overviews provide a high degree of interactivity, very responsive web user interface. An example of web user interface is presented on Fig. 1.

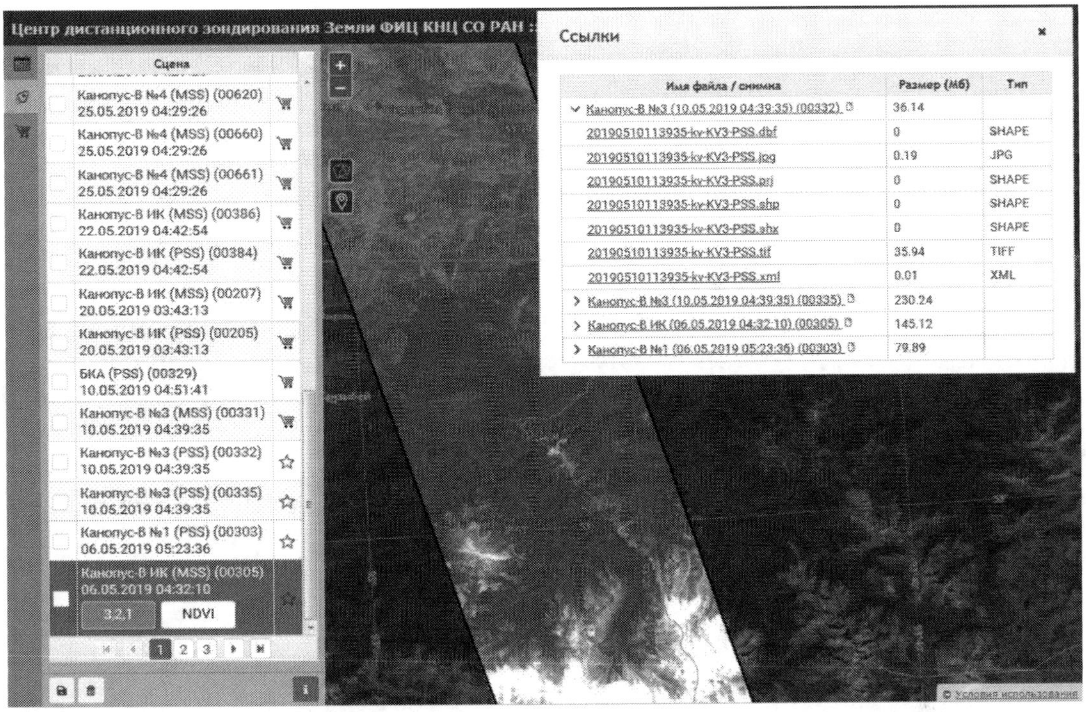


**Fig. 1.** Web interface of the Cataloging system for Russian satellites remote sensing data.

For multispectral satellite images Canopus-V, when creating overview images, a combination of 3-2-1 bands is used, for multichannel images, the following sets of channels are used: 3-2-1 natural colors; 4 and 3 for calculating the NDVI index overview images.

For multispectral satellite images of the Geoton device of the Resurs-P satellite, when creating overview images, a combination of 4-3-1 channels for multichannel images and one channel for a panchromatic spectrum is used.

When forming color images, LUT (Look Up Table) is used, this is a kind of "correction table" for making changes to each of the three channels. Previously, to improve the quality of the image, spectral transformation was used, which is based on working with a spectral diagram showing the relationship between the number of image pixels and the values of spectral brightness. Spectral transformations change such a parameter as contrast. To increase the contrast, a linear stretching of the histogram was used, which means that all brightness values are assigned new values in order to cover the entire possible range from 0 to 255. Applying a common LUT to all images allows you to change the brightness value of image pixels using linear interpolation between, allows you to make images brighter and comparable in contrast to each other. At the same time, the tables allow converting the pixel value of images to 8 bits from 16 bits.

As a result of the work for a quick and convenient search in the satellite data catalog, to minimize the load on the server software and hardware, a set of server applications was prepared for preliminary processing of satellite data, including the following processing steps:

1. Converting the original raster data to GeoTIFF format with converting the original projection to the azimuthal Lambert equal-area projection which is the standard on the regional geoportal.

2. Creation of a basic color image for different display scales, consisting of several spectral channels. Such images will be used for detailed viewing of a satellite image while maintaining the original resolution.

3. Creation of a raster image in PNG format for displaying quicklook in a web application for different display scales, but with a lower resolution (up to 1 km per point). Unlike the basic color image, quicklook in a web application can be displayed in groups depending on the user's choice, while the basic color image will only be active for one selected snapshot.

4. Form of raster multichannel images in GeoTIFF format with NDVI data.

5. Conversion from 16-bit format to 8-bit, which is more suitable for displaying satellite images in a web application and requires significantly less resources for the data storage.

It should be noted that two data formats are used: TIF and PNG. The fast data view is implemented using client-side Javascript using PNG image files. In turn, collections of multichannel overview TIF files of different sizes are used by the web application on the server side. The raster data required for the client part of the software is dynamically created by the server components of the developed system.

An example of the developed web interface of the Cataloging system for Russian satellites remote sensing data is shown in Fig. 1. It also includes the set of links for selected data ftp download. Additional information such as a standardized metadata file and a satellite image outline in the popular GIS file format (SHP-file) can also be downloaded.

## 3 Results and discussion

In 2019, the Krasnoyarsk Scientific Center placed an order with Roskosmos for images from April to October 2019 inclusively from the Canopus spacecraft of the PSS (panchromatic imaging system) and MSS (multispectral imaging system) equipment with a resolution of 2.7 m and 12 m, respectively [6]. And also, an order from the Resurs-P apparatus of the Sangur-1U equipment with panchromatic and spectral imaging with a resolution of 1 and 3 m, respectively. Data were requested with processing levels 1A and 2A with cloud cover no more than 40%.

Areas of interest are shown in Figure 2. Areas of interest include areas in the Krasnoyarsk Territory, the Republic of Sakha, Novosibirsk, Kemerovo, Irkutsk Regions, the Republic of Buryatia and Tyva. In total, within the framework of the submitted order, 31 sites with a total area of about 690 thousand square kilometers were formed.

Data for 2019 has been uneven and the latest data for 2019 was obtained in March 2020. A total of 1103 images were obtained from the Canopus apparatus and 110 images from the Resurs-P apparatus. The total number did not include images with damaged or missing raster files in the amount of 7 pieces.

The disadvantages include the small size of some images, which are located in the corners of the order sites, mainly for images from Canopus vehicles. The size of such images may not exceed 10% of the total area of the site and there is little practical purpose in them. For small areas during the year from the Resurs-P apparatus, on average, one or two multispectral images were collected and usually these images do not completely cover the area. For large areas, the maximum number of images reached 6 and they also do not completely cover the territory. It is impossible to observe the dynamics during the year according to the data from the Resurs-P apparatus. With the data of the Canopus apparatus, the situation is slightly better, but not in all areas.

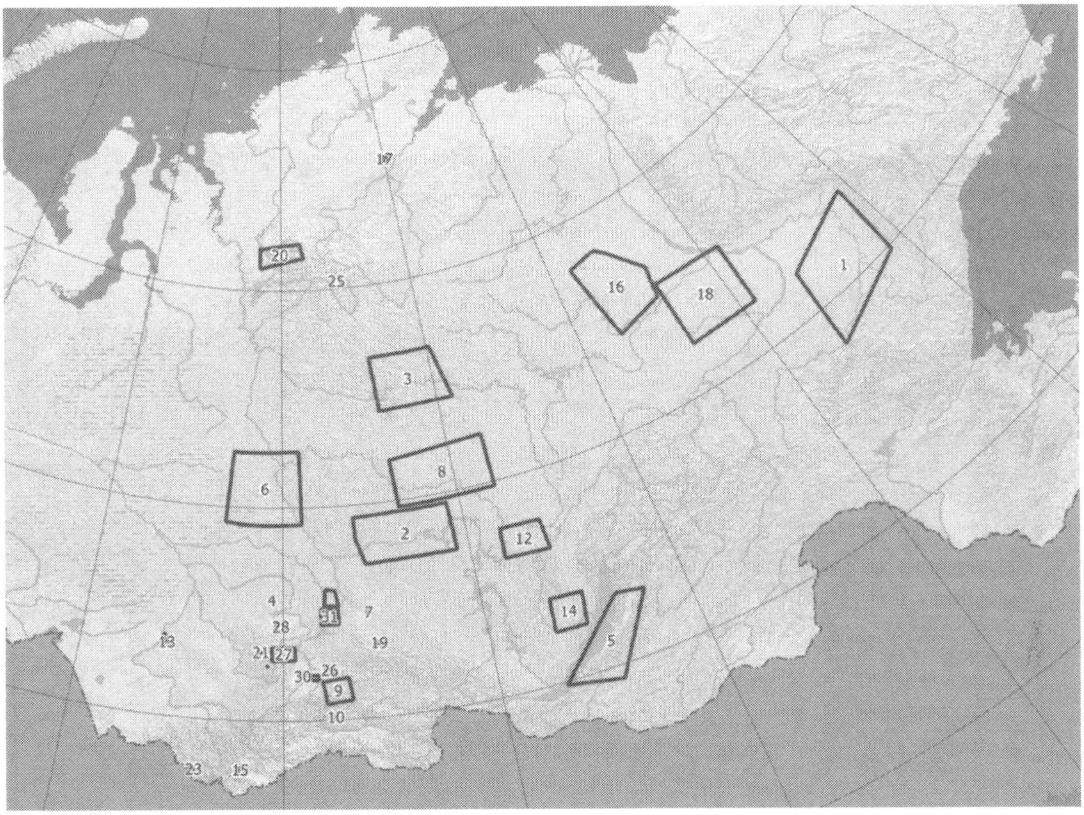


**Fig. 2.** Areas of interest for ordering satellite imagery in Siberia and the Far East.

To provide all interested users within the FRC KSC SB RAS, a centralized archive of all received images from the Roscosmos geoportal was created. A complex of programs was developed for semi-automatic processing of the obtained data. Additional semi-automatic processing is associated with bringing the catalog of images to a unified format with the solution of all the problems described above. As a result, the image file names contain information about the time the image was taken, the type (panchromatic or multispectral survey) and the identifier of the satellite equipment. The hierarchy of image catalogs on the first level contains the shooting month, on the second level the images themselves are located, where the files of each image are in a separate folder. Additionally, for the convenience of users, there is a hierarchy of order sites with symbolic links to folders with images. The developed system processes the data automatically, but in situations where it is impossible to do it, it sends a message to the system administrator for making a decision and manual correction.

To search and navigate the satellite data catalog, ftp access was organized within the network of the Federal Research Center of the KSC SB RAS and a web application for the satellite images catalog was developed using the server programming language PHP, PostgreSQL DBMS, SQLite3 and the client web interface in the TypeScript programming language using the Angular framework 5 [7, 8].

## 4 Conclusions

The creation of an effective software and technological toolkit for the tasks of regional satellite monitoring based on technologies of geoinformation web systems has significant prospects. The developed approach can become the basis for solving urgent applied problems, based on the use of remote sensing data of the Earth, modern spatial data infrastructure, for effective socio-economic and innovative development, increasing competitiveness and ensuring safe life.

## References

1. Kashnitskii, A.V., Lupyan, E.A., Balashov, I.V., Konstantinova, A.M. Atmospheric and Oceanic Optics, **30** (1), 84 (2017)
2. Pinde Fu, Jiulin Sun *Web GIS: principles and applications*, ESRI Press, 2011.
3. Li Songnian, Suzana Dragicevic, Bert Veenendaal. *Advances in Web-based GIS, Mapping Services and Applications*, CRC Press, 2011.
4. J.T. Sample, K. Shaw, S. Tu, et al *Geospatial Services and Applications for the Internet*, Springer-Verlag US, 2008.
5. Mari R., Bottai L, Busillo C, Calastrini F. et al. Renewable Energy, **36**, 754 (2011)
6. Kadochnikov A.A. Geodezia i Kartografia, **80** (1), 110 (2019)
7. Sharma V. K., Amminedu E. et al. Annals of GIS. **23** (1), 55 (2017)
8. Yuqi Bai, Liping Di. Computers & Geosciences. **37** (4), 435 (2011)

# AUTHOR INDEX